C0-AQZ-993

The Role of Contract Engineering in Technical Manpower Planning

The Role of Contract Engineering in Technical Manpower Planning

Paul C. Ergler

American Management Association, Inc.

This book has been distributed without charge to AMA members enrolled in the R&D Division.

International standard book number: 0–8144–2143–1
Library of Congress catalog card number: 70–131551

Foreword

In dealing with problems concerned with technical manpower planning and staffing, engineering managers may overlook the solutions provided by contract engineering (that is, rented talent), perhaps because they are unfamiliar with the concept. However, in the recent past, rented technical talent has been used with success more and more by almost all technologically oriented industries.

The engineering manager, after identifying his manpower requirements, will want to explore alternative approaches to staffing, approaches directed toward satisfying his specific needs and based, where possible, on comparative evaluation. The purpose of this book is to provide such a basis for evaluation; to present contract engineering's role in manpower planning in such a way that the engineering manager may be provided with insight into the concept, the benefits of its use, and its potential pitfalls. To accomplish this end, the book will present the results of surveys (generally through personal interviews) of users and suppliers of rented technical services in the Baltimore, Maryland–Washington, D.C., area. All the material is based upon the author's dissertation research, completed in 1969 at the George Washington University.

PAUL C. ERGLER

Contents

I

Contract Engineering and Its Alternatives

CONTRACT engineering is the renting of technical talent to an organization that performs engineering or scientific work and wishes to augment its staff without increasing the number of its permanent employees. It is a controversial staffing method, because of the rental rates charged to clients and the interpersonal relationships that develop between permanent and temporary employees; some organizations use it only reluctantly, while others do not use it at all—even when it would provide the optimal answer to a specific manpower problem.

Contract engineering has been known for many years as "job shop" engineering, or, more recently, as "contract technical services." The contract technical services industry, which is a part of a larger group called the temporary help services (THS) industry, represents professional services, as opposed to industrial labor or clerical office services. The number of firms offering either clerical or technical services has grown rapidly since World War II, but it was not until recently that the two fields were brought together in one company. Now a few of the larger THS firms are diversifying. Manpower Incorporated and Kelly Services, for instance, have taken advantage of large numbers of privately owned and franchised offices and have started technical divisions. Conversely, some of the larger technical firms, like Consultants and Designers, Incorporated, have expanded into the office services business. It is currently feasible, therefore, to utilize one THS firm for all rented manpower skills that an industrial, construction, or governmental client may require. This is particularly advantageous to the client, because it minimizes clerical work by requiring payment for only one billing for all temporary services.

How It All Began

Although the basic concept of contract help has been used in the construction industry since ancient times, the first modern use is generally credited to the automotive industry prior to World War II. During the 1930s, the industry experienced long slack periods in its engineering work and preferred not to maintain a permanent, highly paid professional staff for peak requirements. The industry has used the contract engineering concept in tooling up for new models ever since. In addition, many overseas U.S. oil company subsidiary plants were built using contract services.

The advent of World War II brought about many skill shortages in the United States, which, together with the new technical manpower needs, caused a number of firms to enter the contract engineering business for a profit. These firms grew rapidly in the postwar years to meet the demands of the expanding electronics industry. Finally, in the 1960s, cyclical conditions arose similar to those encountered in the 1930s by the automotive industry. This time, the industries affected most were those whose major customer was the government. In this case, cost-plus contracts were replaced with more efficiency-motivating contracts, and corporations could not afford to keep idle engineers on their payrolls in anticipation of future work or in anticipation of manning the peak requirements of existing work plans.

From an economic viewpoint, the growth of the technical services industry since World War II is a result of the rapid rate of technological change and development that has occurred and is still occurring in the business and government sectors of the community. Much of this change, in turn, has been an outgrowth of the consuming public's desire for higher living standards in the postwar economic boom. With the increase of competition for similar products in the marketplace, the demand for engineers and technical personnel has continually outstripped the supply, except in certain local situations.

Product innovation is now a key to market share retention and growth. The consequent shortening of new product cycles has required a greater expenditure of engineering man-hours—some of it erratic—for the achievement of profit goals. The engineering manager has been turning to the technical services industry for help at an increasing rate, and will probably continue to do so.

The THS industry is composed of national, regional, and local organizations, although the vast majority of both technical and traditional firms are local and privately owned. Many more technical than traditional firms are nationally and regionally oriented, in large measure because of the perennial shortage of engineers and scientists and the spread of government defense or aerospace contracts across the country.

Margaret Pacey, reporting in *Barron's*, estimated that as of March 1968 there were approximately 1,000 firms in the THS business, half of which supplied technical personnel. In 1967 these firms topped $1 billion in sales, with half going to the technical group. This is compared with $200 million in sales for the industry as a whole in 1960—an estimated growth of 400 percent in seven years.[1] Ernest J. Milani, president of Lehigh Design Company, Incorporated, estimated contract engineering

[1] Margaret D. Pacey, "Money Making Form: Providers of Temporary Help Are Working to Regain It," *Barron's*, March 4, 1968, p. 11.

sales for the 500 contract engineering firms in 1967 to be $400 million, as compared with the $75 million figure in 1957.[2]

The National Technical Services Association (NTSA), which was first organized in 1966, conducted a survey in 1967 of member and nonmember contract engineering firms. Of the 407 mailed questionnaires, they received 105 replies. The replies indicated that dollar sales for the fiscal year ending nearest to December 1966 amounted to $408,024,000. The same survey reported that those in business for the fiscal years ending nearest to December 1965 and December 1961 did business amounting to $237,045,000 and $111,600,000, respectively. If these data are correct, the one-year span (1965–66) produced a growth in sales of 72 percent and the five-year span (1961–66), 266 percent.[3]

Margaret Pacey noted in 1968 that the NTSA claimed a total of 450 firms in the technical field, 82 of which were member firms. These 450 firms grossed from $800 million to $1 billion in total sales in 1967, while the 82 firms grossed $450 million of the total amount. It was suggested that a third of these values included in-house work, which is not normally classified as job shopping.[4]

Who Uses It

Contract engineering firms (or job shops) serve manufacturing industries, construction industries, research and development organizations, and government. They are most prevalent in fields that are characterized by employment instability or abrupt changes in staff requirements, usually caused by model changes, technological breakthroughs, seasonal changes, business cycle fluctuations, and the like.

[2] Ernest J. Milani, "The Contract Engineer," *Mechanical Engineering*, September 1967, pp. 52–53.

[3] Letter from Sheldon J. Hauck, Executive Director, National Technical Services Association, Washington, D.C., February 6, 1968 (including report of 1967 survey dated October 1967).

[4] "Money Making Form," p. 11.

Because aerospace and other defense-oriented industries during and after World War II contributed in large measure to the growth of technical services suppliers, many potential users today associate job shopping with these industries to the exclusion of all others. However, the concept is also used extensively in the automotive and electronics industries, data processing, tool and machinery design, chemical and petroleum plant construction, and public utilities.

The author conducted a survey in the Baltimore, Maryland–Washington, D.C., area, questioning 32 prominent supplier firms out of the 47 companies that supply technical talent to the area. One of the results of the survey included a distribution of job shop employees according to the industry to which they were rented. These results are as follows:

Industry	*Percent of Total Shoppers*
Electronics	34.1
Aerospace	21.3
U.S. government	11.8
Mechanical machinery and heavy industries	10.5
Engineering and architecture	5.3
Electrical machinery and utilities	5.2
Business service	3.8
Chemical	3.3
Ordnance	2.3
All others	2.4
	100.0

It was estimated from these results that 50 percent of the job shoppers in the area were engaged in U.S. government (or government-sponsored) work, while the remaining 50 percent were distributed in varying proportions among a variety of industries in nongovernment, nondefense, and nonaerospace efforts. However, many other industries are now using the concept, and this trend toward diversification is expected to continue as potential users realize the benefits available to them.

The spectrum of technical services offered by suppliers

ranges from the elite status of professional consulting by licensed professional engineers to the lowest status of contract technician. Some technical services firms have offered the services of contract machinists and assemblers, but because this may invite criticism and retaliation by organized labor unions, machinists and assemblers are not normally considered part of the personnel provided by a technical services firm. The most typical technical services offered are the following:

- Engineering consulting.
- Engineering analysis.
- Engineering design.
- Engineering evaluation (test).
- Engineering drafting and checking.
- Tracing.
- Technical publishing: writing, editing, and illustrating.
- Laboratory and R&D technician service.

In addition, the services of mathematicians, physicists, chemists, geologists, human factors specialists, programmers, electronic data processing analysts, administrators, and many others are offered to clients, depending upon the business emphasis of the technical services firm.

Manpower Planning and Staffing

Staffing, one of the major functions of management, is dependent for its effectiveness upon a consideration of other functions. This is particularly true of the advance planning for required manpower in simple as well as complex situations. If the planning is inadequate and the resultant staffing reflects poor timing and improper skill levels or mixes, overall performance as measured by costs and schedule may well be unacceptable. Since management must anticipate continual and sometimes unexpected changes in the process of striving for goals, there must be a certain amount of staffing flexibility to permit rapid increases or decreases in specific types of personnel, when necessary, and to give management optimal control of manpower expenditures.

Staffing flexibility, however, is difficult to achieve. Humanitarian considerations and human inertia reduce the willingness to change, even when change is necessary—particularly when it is necessary to revise skill mixes and levels or to destaff. The tendency is to "make do" with the present staff or to avoid layoffs until the situation becomes critical. Conversely, when rapid additions to manpower are required, inertia again thwarts accomplishment; it takes time to hire capable personnel who are already working elsewhere.

It is important to solve staffing problems in any managerial situation, and most important of all for the engineering manager in technologically advanced industries. Skill level, skill mix, and cost all tend to be more critical in these industries, because of the number of varied and expensive personnel with whom the technical manager must work. Furthermore, the total engineering manpower that is required as a function of the time schedule for various skills tends to oscillate in more pronounced cycles than it does in less technologically oriented industries.

Manpower Requirements

Manpower planning for an engineering program should establish the detailed effort in terms of engineering man-hours for each major task. This must be done in conjunction with the master schedule plan for the accomplishment of specific goals within the overall plan. Preliminary Gantt charts are prepared; they in turn are used to assist in the preparation of PERT diagrams; and the individual Gantt charts and PERT diagrams are then coordinated and revised until integrated planning is effected. Summations are made of the required technical manpower according to skill, task, and total program, and these requirement summations are then included in the total summations for the plant so that detailed staffing plans can be formulated. These total curves should include all in-house projects that are firm contracts together with those that have a high probability of becoming firm. An illustration of the planned total program and plant technical manpower requirements for a hypothetical high-technology-oriented company is presented in Exhibit 1.

The figure in this exhibit represents the planned manpower required to accomplish the various phases of a program. Other programs begun during the same time period will increase the technical manpower needed by the entire company. The ordinary time sequence of technical activity for a program designated Model X may incorporate new research and development results that can enhance proposals submitted to the customer or to top management. After successful competition for available contracts, or after a decision by management to proceed, the prototype hardware article is designed, built, and tested. If the nature of the product leads to production quantities of Model X, an engineering activity aimed at producibility and reliability is initiated. Following this, a minimum continual change and maintenance effort is required for as long as production is in effect.

The major staffing problem that faces engineering management is to fit the proper numbers of people with the proper skills to the required manpower curve as a function of time. If actual manpower expenditures do not fit the requirements, inefficiencies will result and schedules may be missed. Several technical staffing tools will assist engineering management in gaining the flexibility that it needs in order to match the manpower requirements curve with expenditures.

Requirement-Matching Considerations

As previously noted, matching manpower requirements with actual staff and then maintaining control of the numbers and skills used are the keys to performance—if the initial homework of planning has been taken seriously. The first step in program

Exhibit 1

Technical Manpower Requirements

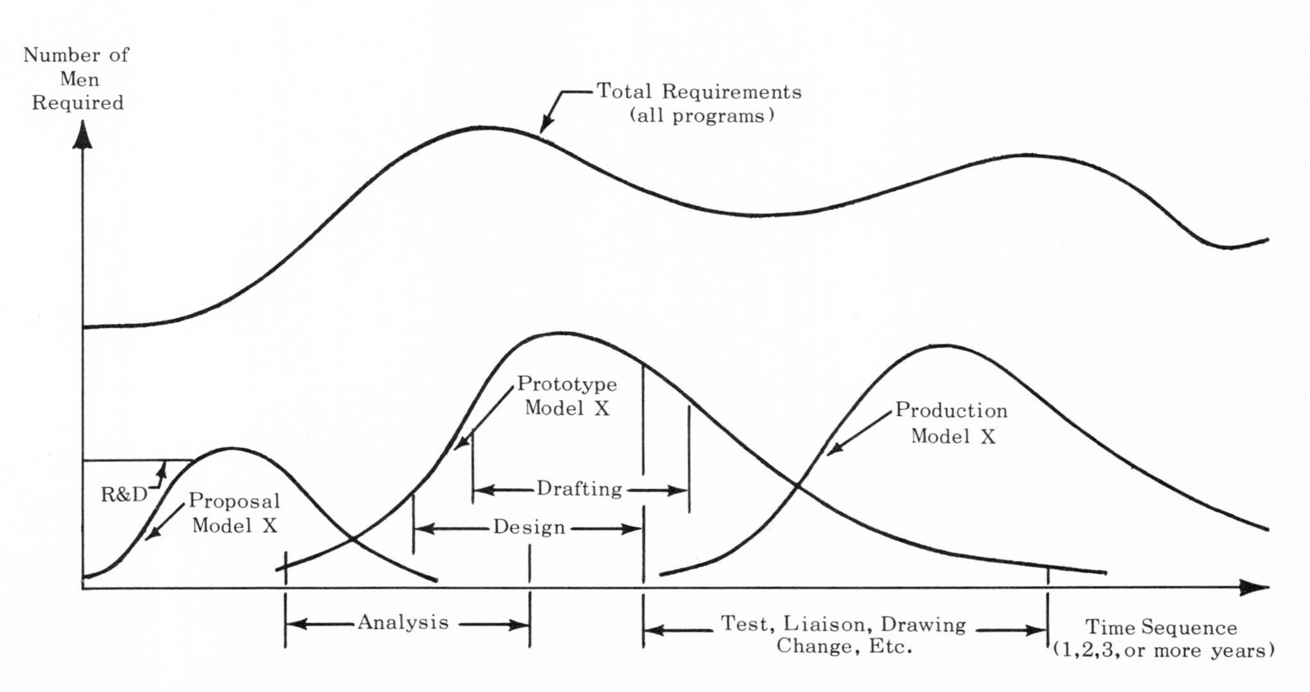

requirement matching is to modify the requirements to permit the maximum effective use of personnel who are already available elsewhere in the organization. Thus the program technical director and the plant engineering director (or their representatives) must ask the following questions and be prepared to seek honest answers:

1. Can other programs be rescheduled so that the immediate demands for manpower on this one can be accommodated with existing staff?
2. Is the existing staff composed of the skill mix that will provide the flexibility to permit a shifting of people to handle the new assignment?
3. Can the program end date or start date be extended without significant penalty?

If potential internal adjustments fail to provide the needed staffing, and other action is required, the following questions should be considered in advance of choosing a method to match requirements:

1. What are the real costs incurred by overtime work?
2. What are the costs involved in recruiting, hiring, and training a new staff of the proper caliber?
3. What are the costs involved in the use of temporary technical services?
4. What needs to be done, if anything, regarding procurement of office space, laboratories, and equipment?
5. What are the special security requirements imposed by the customer and/or management? How will security or proprietary considerations affect the flexibility of choice, facilities, and so on?

When an engineering manager or a program technical director is required to increase his engineering effort and plans to do so according to requirements similar to those described in

Exhibit 1, he must decide how he can most effectively meet the manpower demands. His alternatives are as follows:

1. Work the staff overtime.
2. Hire new personnel or transfer employees from other divisions to increase the permanent staff.
3. Subcontract a portion of the workload to a technical service company, a consulting engineering firm, another similar company with experience and know-how in the field, or to another division of the parent company capable of handling the work.
4. Hire or contract for temporary help to supplement the existing staff.

The manager's decision as to which alternative or combination of alternatives to choose must be based on an analysis of the problems at hand and on answers to the questions presented in the two previous lists. He must think in terms of his schedule, the cost of his decision, and the feasibility of such a decision within his overall plan.

Overtime work. Before deciding to work the staff overtime, the manager should determine whether the men are willing, how the overtime work will affect their morale, and how the predicted state of morale will affect job efficiency. Will personnel be dissatisfied with their income when overtime is terminated? If management demands unpaid overtime on a scheduled basis, will permanent employees tend to seek other employment?

Staffing by permanent hire. Before deciding to hire and increase the permanent staff, the manager should determine whether the program can wait until the organization can recruit, hire, and train the new people. If the skills are in nationwide demand, it may take from six months to a year to put some of them on the payroll. A very special skill may take longer. The recruiting procedure not only takes time but can involve a cost of thousands of dollars per hire.

Alexander Terentiev, a technical manager of Comprehensive

Designers, Incorporated, has cited a cost estimate for new hires:

> Each new engineer costs a prime company about $25,000 in overhead. These costs include recruiting expense, training, benefits, and other overhead items such as non-production during slow periods.[5]

Charles E. Zimmerman, president of Consultants and Designers, Incorporated, whose company is a leader in supplying technical temporary help, is of the same opinion as Mr. Terentiev and has stated that: "Every new technical employee permanently hired by a company costs that company from 100 to 200 percent more than his basic salary."[6] Mr. Zimmerman includes in his estimate the costs of recruiting, relocating, training, fringe benefits, social security and other taxes, vacation and sick pay, jury duty pay, and other overhead expenses. Stephen N. Elias, vice-president of ARCS Industries, believes that a company must have two years to amortize the overhead of a payroll employee before he becomes competitive with a temporary help acquisition.[7]

It is pertinent in increasing the permanent staff to consider the recall of laid-off employees, if any, and to utilize other departments and divisions of the company for transfer of the skills that are needed or that could be converted to needs through a training or professional development program. This, from a good personnel relations standpoint, should precede the hiring of permanent or temporary help.

Permanent employee hiring involves a costly initial program with many residual (or hidden) costs that may be overlooked. Recruiting demands advertising in various population centers and in trade magazines to attract the required talents. If the market is tight or a skill scarce, it may be necessary to engage

[5] Martin Gold, "Job Shop Revisited," *Electronic News,* July 18, 1966.

[6] David Fouquet, "New Job Gypsies Boast Skill, Top Pay," *The Washington Post,* May 1, 1966, p. L1.

[7] "Job Shops: They're Big Business," *Engineering News Record,* August 10, 1967, p. 16.

a personnel agency to supply screened résumés and later send the candidate for interview. Company funds must often be expended on attendance at national meetings of professional engineering societies—including the use of "hospitality rooms" in hotels and the use of top level company engineers as temporary recruiters—in order to acquire the highest quality of technical employees. Then there are plant visitation expenses and interruptions of operations for tours and more interviews. After all these recruiting expenditures, only a small percentage of applicants are usually acceptable, and a much smaller percentage accept the offer of employment. And even then, there are still more expenses: charges for travel, moving, per diem living expenses, orientation training, and adjustment inefficiencies on the part of the new hire and all with whom he interfaces.

One important consideration in the hiring of the required technical talent is the assurance that the new hire will not need to be terminated within a forecast period of at least one year because there is no work for him. Such terminations create poor community relations, represent poor business ethics, and are upsetting to longer-service personnel who observe that their security is threatened.

Subcontracting. If the engineering manager decides to subcontract a portion of his workload to outside organizations, he should make an advance study of the procurement of goods and services and the attendant applicable arts used in his organization. The portion of work taken from the whole must be capable of easy division with a minimum of interface problems; otherwise, coordination efforts will be complex and costly—particularly if the work is to be done in a geographically remote location.

The subcontracting philosophy can be oriented in one of several directions. The subcontracted portion can be a subsystem or component of the total job and still be reasonably independent of the rest of the work effort. On the other hand, technical work that is integrated with the rest of the effort can also be subcontracted. The former method requires the subcontractor to apply top talent at many skill levels, as in the detail design of an outer wing panel for an aircraft, with all associated documentation as the delivered items. An example of the latter method might be the preparation of maintenance and service manuals, which must be written, illustrated, edited, and published—requiring many inputs and much coordination from the prime contractor's staff in all skill areas.

These types of subcontract work can be handled by technical services organizations on their own premises. The more complicated jobs can also be subcontracted to a competitor or to a remote division of the same company that may be able to handle the workload. Further, if the type of work is applicable, conventional consulting engineering firms can be invited to bid. A major advantage of subcontracting is that the facilities and equipment of another firm are used; the capital saved may be applied elsewhere in the business.

Staffing by temporary hire. Before the engineering manager makes the decision to use contract engineering help (commonly known as job shoppers) on his own premises, he should consider a number of temporary help approaches:

1. Pensioned employees who possess the required skills can be called back. They can be hired as consultants and require little or no indoctrination. An up-to-date listing of those who are willing and able to work should be maintained.

2. Companies that belong to industry associations can encourage the formation of groups or committees to help each other survive the fluctuating requirements of modern engineering department manpower cycles. Engineering administrators of certain aerospace companies, for instance, meet periodically to discuss common problems. Many times their problems of manpower shortage and surplus have been solved by talent borrowed from one another. Currently, this approach is also effective among professional consulting engineering firms.

3. It may be possible to hire former employees on a "moonlight" basis if the man-hour requirements and time limitations permit. The advantage in doing this, of course, is to achieve immediate effectiveness without training or delay.

4. It may be possible to fill part or all of the requirements

by using faculty and students of various colleges of engineering. Although most students tend to be available only in the summer, cooperative education students are available for short-term employment all year long (from three to six months at a time). There are a number of colleges offering five-year engineering programs that have these cooperative education departments.

One would surmise that college faculty, too, would be available only in the summer; actually, many are available as consultants and part-time employees all year long. Furthermore, if the school is on a quarter system rather than a semester system, certain faculty members can be made available according to advance planning for a three-month, full-time period in the fall, winter, spring, or summer term. In essence, they become freelance technical temporaries and bring modern academic talent to bear on a technological problem as required by the program schedule. This idea was advanced by LeRoy A. Brothers, then Dean of Engineering, Drexel Institute of Technology, who expressed interest in periodic industrial experience for his engineering staff.[8]

5. If any of these four temporary-help approaches (or indeed any of the nontemporary-help concepts previously discussed) fails to provide adequate solutions to the staffing problem, it may be desirable to investigate the use of a technical services firm for contract engineering or job shop talent. Furthermore, if other solutions appear adequate, overlooking the rental of talent as an alternative for comparison purposes may provide less than an optimum solution to a particular problem.

So that the engineering manager may be able to evaluate more fully the contract engineering concept as a possible alternative in his staffing decision, the following chapters of this book will outline the concept in greater detail and define its present role, its future direction, and the advantages and limitations involved in its use.

[8] Interview with LeRoy A. Brothers, Provost, Drexel University, Philadelphia, Pennsylvania, June 9, 1968.

2

The Advantages of Contract Engineering

THE role of contract engineering can best be described and illustrated through the actual reasons given by user organizations for their acceptance of it. Hence in a survey of 45 user organizations of various industries and sizes (19 large, 12 medium, and 14 small, where "large" is defined as having gross annual sales of over $50 million and "small" is defined as having gross annual sales of less than $10 million), engineering managers were asked to state the reasons for their use of leased technical people. Their answers can be summarized as follows:

Reasons for Use	*Percent of Sample*
1. Meet peak requirements	98
2. Economical	29
3. Talent more effective than own	16
4. Can't get new hires quickly	13
5. Prevent missed schedule penalties	11

	Percent of Sample
6. Policy to prevent layoff	7
7. Flexibility and convenience	7
8. Company geographic move announcement	2
9. Layoff callback list exhausted	2

Of the nine categories of reasons given by the users, not only was "to meet peak manpower requirements" stated in 98 percent of the interviews, but several of the other categories were obviously related to it, such as "to prevent missed schedule penalties" and "to prevent layoff."

Almost a third of the sampled users stated that contract engineering was economical for them for many different reasons. In some companies, the breakeven point for economical considerations was eight or nine months; that is, if the personnel were needed for less than this time (the average leasing period was five months), it was more economical to use shoppers. Other companies realized economy through the increased efficiency of captive personnel when temporaries were used, for example, when overtime was eliminated and captives could return to a normal work week. Defense firms indicated that the use of contract engineering was advantageous for them in relation to overhead allocations.

Many organizations found that they could obtain better or more effective talent by renting than they could by hiring or using their own staffs. One company pointed out that it could rent an older and more experienced man for a specific job than it could ever expect, need, or be able to hire on a permanent basis. On some occasions the required talent was not available within the company.

According to the detailed data on which the summary of reasons was based, three reasons for use may have been influenced by the size of the user company. The larger companies seemed to have a more difficult time getting people quickly when they were needed; the smaller companies did not seem to be "schedule conscious"; and the medium-size companies emphasized the use of a temporary-help, no-hire, no-fire policy.

"Reasons for use" and "advantages of use" are almost synonymous—but not quite. The *reasons* shown here were presented by a sample of users and represent actual situations. However, determining the *advantages* is the result of creative thinking about the concept. Thus even if any advantages mentioned in the following discussions have not also been noted as reasons in the survey, they may still become reasons in some future use of the concept.

1. The most impressive advantage to the user of the service is that he can get expert, experienced help in a hurry in order to meet peak requirements. Most of the time, the necessary talent can be obtained in a few days. Depending upon the volume of people required, the company requiring the help often needs to call only one job shop (or at most several), with no costly advertising or personnel campaigns and no bargaining with individual people regarding salaries and benefits. If a client company uses a reputable job shop in which it has confidence, the operation of acquiring the needed help can be efficient.

Mr. Ernest Milani, president of Lehigh Design Company, reported a dramatic example of the speed at which firms are sometimes called upon to perform. He said: "In New York, one Friday afternoon at 4:45 P.M., our firm received an order for 135 subprofessionals to be at an out-of-town job the following Monday at 8:00 A.M. The order was filled."[1]

Mr. Milani does not recommend giving such short notice habitually, since it prevents the technical service firm from doing the thorough screening job demanded for assurance of quality. But when crash programs materialize and large scale technical staffing is needed at once, no other form of staffing may be available.

2. A company can have the flexibility to expand and reduce technical manpower to meet peak load periods in the engineering schedule. If the company is small, it can hire permanent draftsmen for general use and rent engineers for specific jobs—or vice versa.

[1] Ernest J. Milani, "The New Breed of Gypsy Technicians," *Personnel,* November–December 1967, p. 60.

3. Specialized and scarce technical skills can be rented for the specific time needed to augment the capabilities of the engineering staff.

4. The captive personnel on the project may benefit from fresh viewpoints that are brought in by outside technical people with a variety of experience.

5. There is a shorter learning time for contract personnel than for new hires.

6. A company can reduce its long-term overhead expenses by not having to maintain certain jobs or skills on a full-time basis. Then, too, the amount of costly overtime work by the captive staff can be reduced. Job shoppers are paid on an "as worked" basis. Thus the user of the service does not have to provide regular employee benefits or payroll taxes. Furthermore, assignments can be expanded or reduced without worry about the shoppers' morale.

7. If permanent employees are told in advance that temporary help will be used, and have the reasons explained, their morale will be improved; they will know that the temporary help will relieve them of excess work at peak load times, assist them with special tasks, and fill in, where possible, during vacations and sick leaves. Furthermore, regular employees can concentrate on their work without the fear of layoff.

8. An advantage that should not be overlooked is the fact that contract firms guarantee satisfaction. If within a stipulated period the client company determines that the rented employee cannot do the job, it can request a replacement at no charge for the lost time. The grace period usually does not exceed a few days; however, even after the grace period ends, the client can request an immediate replacement or return the temporary employee without replacement at any time. Furthermore, shoppers can be interviewed prior to acceptance on the job.

9. The need to hire, train, and later terminate people on the regular staff after the peak period of the project can be minimized. If such terminations are frequent, a company's employee relations and recruiting programs can be damaged. Furthermore, the entire community may express its resentment at poor

personnel practices. Job shopping permits a stable permanent staff to be maintained on the basis of minimal needs.

10. The client company of a technical service firm has the advantage and opportunity of observing job shoppers for possible permanent employment. The shopper, in turn, may actually be determining from short-term experience just where and with what company he would like to settle down as a permanent employee. Here is an opportunity to hire new people with full assurance that they are able to perform as desired.

Most contract firms specify in the contract a minimum period of about three months before a switch to permanent status can occur without a penalty. When and if the switch does occur, the client company does not have to pay employment agency fees, which usually amount to 5–10 percent of the hire's annual salary.

11. The advantage of job shopping for the client company that has moved its plant to a remote geographical location (usually to benefit from low taxes and low-cost production workers) is obvious. Many engineers prefer metropolitan centers where opportunities for graduate study and cultural benefits for families are available. Subprofessionals such as the designer, draftsman, and technician may be interested in university course work to stay up to date or to complete a bachelor's degree. The remote firm requiring technical talent has a problem that can be solved by job shopping. In such cases the terms of employment for contract engineers are usually long and the rates are premium—but the expenditure is worthwhile in terms of accomplishment.

12. Another use for the contract engineer is in securing contracts. These situations are prevalent in the aerospace and defense industries where several competitors may be awaiting the results of a competitive evaluation. In this case, the schedule commitment may be such that a staff must, in effect, already be working, in order to accomplish such things as subsystem and component subcontract awards early enough to give the potential subcontractor his required lead time. If a company fails to

win a contract, it can destaff immediately by removing the temporary contract engineers.

Some contract firms will enter into a "partnership" with small companies and help them bid on large contracts, with the provision that the helping firm will be retained if the award is won. It is also currently acceptable, in bidding on government contracts as potential prime contractors, to state how the bidder plans to staff the engineering effort. As such, many bidders now list an affiliation with one or more specific job shops, as assurance that the work can be staffed with the needed manpower in time to deliver the product.

13. The temporary services of technical people can be used to advantage when a high-risk special program is pursued. Frank H. Rumpf, president of the Institute of Temporary Services, noted that

> R&D firms requiring whole teams of designers and scientists to pursue a speculative new idea welcome the temporary services as a source of talent without suffering the penalties and dislocations which stem from large lay-offs when a project is abandoned or completed.[2]

It is thus seen that contract engineering is a handy tool for commercial firms that are trying to surpass competition by innovating, shortening new product cycles, and gambling that their ideas will be a success. Mr. Rumpf also noted that top executives and whole teams have been used to manage and man special projects on a temporary basis.

14. The use of temporary help can serve to relieve recruiting pressures and allow personnel people to concentrate on hiring quality permanent people. Urgent demands for hired help can reduce the effective search for permanent personnel; therefore, it is an advantage to hire temporary personnel while the search is being conducted.

[2] Frank H. Rumpf, "How Temporary Help Boosts Our Economy," *Office*, January 1967, p. 92.

15. Contract engineering permits the nation to make the best use of its technical resources by moving highly skilled individuals to the geographical areas in which they are needed.

Empirical Indications of Satisfaction

Organizations hesitating to use job shoppers may be moved toward consideration of the method if they can observe some amount of general satisfaction with it on the part of a representative sample of previous users. One indication of satisfaction is the repetitive use of individual job shoppers. Let us assume that a project manpower peak within a company has passed, and the job shoppers have been released. Now, if another large manpower requirement becomes imminent, it will prove significant if engineering management requests the same individual job shoppers to return. It is the author's belief, on the basis of his engineering management experience, that if one-third of the job shoppers are requested to return, the clients have been receiving reasonably good quality and the service has been accomplishing its purpose.

In a survey by the author, representative samples of both suppliers and clients were asked to provide data on callback requests by individual name. The two sources provided a validity check on each other and indicated whether the suppliers, as salesmen, were biased upward in their perception of user satisfaction.

Analysis of the observed data indicated that the samples were probably drawn from populations with 50 percent callback requests for individual shoppers, where a statistical level of significance was 2.5 percent. A statistical comparison of data from the two sources (suppliers and clients) revealed that the populations were equal and that any bias offered by suppliers was negligible.

Additional analysis was performed to determine whether callback proportions were affected by the size of the user company. Here again, samples from all three company-size categories were determined to be from equivalent populations with equivalent variances and means, indicating that size did not influence the callback requests. The significance of these results is that the users of contract technical services were so well satisfied with the temporary help they received—at least in the geographical area represented by the survey—that they invited 50 percent of their temporary help to return for another program. Thus they overtly displayed their satisfaction with the concept, despite some negative comments also obtained from them.

Contract Engineering in Reverse

Contract engineering in reverse is a little-used technique for destaffing. However, it has the potential of stabilizing professional staffs, as does the basic concept of renting talent for peak loads. A similar approach was briefly discussed under methods of staffing by temporary hire, where industry associations loan one another technical talent to help survive fluctuating requirements and maintain a stable level of employment.

The concept of job shopping one's own permanent staff for hire puts conventional product-oriented firms in the temporary personnel business on a temporary basis. But most firms are neither knowledgeable about nor comfortable with situations involving manpower marketing, and consequently they may choose to lay off when requirements drop below normal, rather than take an approach that requires innovation.

An example of innovative thinking is the advertisements placed by The Boeing Aircraft Company in *The Wall Street Journal* during the winter of 1970. These advertisements offered to rent all types of engineering and scientific skills—including complete engineering teams—to anyone interested. If the offering were successful, large numbers of people would be saved from being laid off following the 1970 NASA and DOD funding cutbacks.

If a company wishes to job shop its own staff during slack times instead of laying off, but has no real desire to become involved in the personnel rental business, it can seek out certain innovative job shop firms. These firms will specifically contract to supply rental technical personnel during peak loads and also to act as an agent to rent out certain permanent employees during slack periods.

The use of contract engineering in reverse to stabilize one's permanent staff is an efficient approach to destaffing. It enables a profit to be realized on personnel who are not immediately needed; it promotes a ready source of manpower for the next "upswing"; and it can give a company a reputation for stable personnel practices.

The infrequent use of contract engineering in reverse was brought to light in the author's survey of organizations that utilize job shopping. Only 9 percent of the sample ever job-shopped its own permanent staff; and these represented only two industries: (1) aerospace and (2) engineering and architectural firms. When the surveyed organizations were asked what company practices were followed when their workload fell below normal permanent staff levels, the following replies were received:

Reply	*Percent of Sample*
Lay off	31
Stimulate personal research	27
Make work	18
Transfer to another division	16
Engage in company-sponsored technical studies	16
Increase proposal and marketing activities	11
Job permanent employees out	9
Obtain more work from parent company	2

Only a few of these approaches provide an obvious, positive gain for the utilizing company, and job shopping of permanent employees is one of them.

A Comparison with Conventional Consulting

A number of contract engineering firms throughout the country include the word "consultants" in their corporate name. Generally, these firms employ a limited number of professional engineers who possess licenses to perform engineering work in their particular states. They do, to a limited extent, impinge upon the market traditionally served by professional engineering consulting firms.

The conventional professionals until recently have cast a leery eye upon contract engineering firms. For instance, The American Society of Civil Engineers in 1958 issued the following policy statement: "Job shops have an unprofessional effect on the professional practice of civil engineering." A committee of the National Society of Professional Engineers in 1967 stated that "Job shops are not in the best interest of clients who purchase these services, the engineering profession, or the general public." The Consulting Engineers' Council president for 1967, Samuel A. Bogen, stated that "Job shops tend to drive up wage rates, lead to personnel raiding, and generally discredit the profession."[3]

In support of the job shops, it is emphasized that most of the work of the scientists and engineers who accomplished NASA's Mercury, Gemini, and Apollo programs—work that pushed back the frontiers of science and engineering—was not done by professionally licensed engineers. Although a few held state licenses, the vast majority were not licensed and many of these were contract engineering personnel.

Donald A. Buzzell, the executive director of the Consulting Engineers Council of the United States of America (CEC) concurred with this view.[4] Mr. Buzzell pointed out that 10 to 15 percent of CEC firms look upon job shops as competitors and have been trying to put through an official council-sponsored policy statement against the shops. However, after numerous

[3] "Job Shops: They're Big Business," p. 17.

[4] CEC represents 2,100 member firms, each averaging 20–25 people.

meetings over the years, the council has yet to agree on a policy statement. Mr. Buzzell said:

> After all, you have got to have something solid to be against something. We couldn't pin it down as to what to be against. The shops are legal and they perform a service which more and more of the CEC membership are actually using. Because a man has a Professional Engineers' license doesn't give him a monopoly on engineering talent or brains. Furthermore, it is not required by law for engineers in most of our big industries to be licensed—therefore, the CEC really can't be against job shops.[5]

The professionals who dislike job shops and are pressuring CEC the most are those members that are engaged in industrial plant design, those that do general engineering, and those that are involved with urban development. The amount of work being taken away from the members by the job shops was unknown to Mr. Buzzell; however, he noted that some job shops are doing complete designs of buildings with integrated systems. In these cases, CEC believes that the client is not as well served by the shops and that some shops are not operating ethically. The shops, on the other hand, say it is in the client's interest to save money by paying on an hourly rate basis rather than the conventional consulting fee basis. Mr. Buzzell contended that each case is an individual case and one should not generalize as to whether the client is or is not well served. When asked about the future of job shopping, Mr. Buzzell expressed the belief that it would continue to grow, since more and more of his own engineering firms were finding the shops economical to use.

The author conducted a survey of 45 out of a population of over 100 organizations in the Baltimore–Washington area that use contract engineering, to determine whether they used contract engineering as a substitute for the services of professional

[5] Interview with Donald A. Buzzell, Executive Director, Consulting Engineers Council of the U.S.A., Washington, D.C., November 25, 1968.

consulting engineering firms. Thirty-two of the 45 answered in the negative; 9 others said they used the services of both types; and 3 declared themselves partial to contract firms, to the exclusion of the professional consultants. These results may indicate that user firms tend not to use contract engineering as a substitute for the more conventional professional services.

3

The Limitations of Contract Engineering

THE role of contract engineering in technical manpower planning can be positive or negative, depending upon its use and application. This chapter will discuss the negative aspects and limitations of its use, so that the engineering manager may make his own evaluation that will conform to reality.

The Disadvantages

A survey of the current literature on the concept will find little mention of the disadvantages involved in contract engineering. Perhaps the novelty of the "gypsy" engineer overshadows negative thinking; or perhaps most authors have a personal stake in the success of the concept. Be that as it may, there are disadvantages and pitfalls, and if they are recognized in advance, they can generally be avoided. The results of the author's survey of users discovered the following opinions:

	Disadvantages	*Percent of Sample*
1.	Lack or loss of specific knowledge	50
2.	Irresponsible or ineffective workers	41
3.	Poor economics	34
4.	Drop in morale of permanent employees	16
5.	Miscellaneous	19

Lack of specific knowledge. Several user organizations complain that job start-up time is excessive with contract engineering, because few engineers are available with specific knowledge of every industry. There can be additional delay when shoppers have to study and learn a specific company's processes, procedures, and standards. The only compensation for this problem is that shoppers tend to profit from their varied experience, and may even acquire and use new knowledge more rapidly than will new hires.

Unless a potential user checks out the reliability of the contract firm he is about to deal with, he may receive poor, uninformed service and become prejudiced toward the concept. Since such a business is easy to start, it has attracted some people with no prior technical experience. Most of these have failed, but not all of them. A few are totally untrained and give the industry as a whole a bad name through their improper selection of talent for specific job requirements. The result to the client firm that accepts such service is spoiled work and increased costs.

It is unfortunate that suspicion must be cast upon résumés in a business like contract engineering that lives by them, but fictional résumés from shoppers are plentiful. The problem, however, is not peculiar to the job shop firm, but plagues the personnel agency as well—particularly the agency that deals in professional and paraprofessional talent.

The most common fiction is a claim to possess specific engineering degrees that the applicant does not have. In a permanent job situation, the chance of being discovered in the long run is high; therefore, the applicant may be more careful about his statements. But in job shop hiring, where the term of employment with one client is short, the already unscrupulous person may be encouraged to be more so because of the smaller chance of being caught.

The job shop personnel manager may not be technically oriented enough to prevent this form of fraud, or he may not be able to interview the applicant in person. Instead, he may have to rely upon the résumé and a phone call when hiring his contract personnel. The truth can be uncovered only through alert interviewing by knowledgeable engineers or by a detailed security checkout of the stated information.

Loss of specific knowledge. Some users have claimed that shoppers are not necessarily as loyal to the client as are permanent employees, and that they readily tend to quit and move on if they are offered more money elsewhere. When this happens, knowledge is lost and a design can be left uncompleted. Furthermore, the greater the market competition is for engineers, the greater chance the contract engineer will have of bidding his rate upward and, in fact, leaving an expensive job half done. Knowledge can also be lost en masse if the company or government agency that uses the service must resort to competitive bidding among several contract engineering suppliers. In these instances, clients may suffer if they have not developed enough permanent key personnel.

One way of handling the problem of the shopper (or permanent employee) who leaves in mid-job is to set up a bonus arrangement that will convince the worker to stay. A bonus may be given, for example, to personnel who stay until a specified date, or until a specified project has been completed.

An outgrowth of the problem of departing personnel is the company's inability to provide adequate manufacturing liaison. By the time the job shopper's designs reach the floor and start to be fabricated and assembled according to plans, the job shopper is seldom available to solve any problems that may arise. (Of course, there is also a chance that the permanent employee may not be available for such liaison, but there is more of a chance that he will be.) Once the job shopper is gone, the benefit of his prior analysis is lost. When unforeseen mistakes show up in production, it is not known just what the engineer had in mind while he was developing the design, or even what design alternatives he already discarded as being unworkable. Therefore, new talent must be assigned to solve the problem, and production inefficiencies result.

Irresponsible workers. Some users believe that job shoppers spend an abnormal amount of time talking on the telephone during working hours to people outside the facility for purposes of personal business. Of course, it is possible that these workers are conducting necessary personal business, such as setting up dental appointments, but some users claim that they are either negotiating for another job or arranging a date.

Other observers say that contract engineers leave their work each night "before the sound of the bell has died away." With immediate design problems to solve and difficult schedules to meet, it bothers some people to see contract personnel drop their pencils and run; this seems to them to be an indication that the shoppers are in the shopping business just for the money and have little job interest. However, these characteristics can often be seen in engineers and other technical people in general. It has been observed that most technical personnel (unless they are absorbed in some aspect of research) tend to leave at the sound of the bell whether they are regular employees or job shoppers. Those who stay on beyond the bell tend to be technical supervisors and others in management.

In addition to these mild indications of shopper irresponsibility, some clients have made more serious accusations. They say that many shoppers are unreliable and irregular in reporting on the job, and that they claim personal business for absences but do not call in advance. Furthermore, they seem to lack interest and motivation. They just put in time during regular hours; work the overtime for extra pay; and then disappear for days.

One disadvantage associated with the nomadic characteristic

of contract engineering is the opportunity for personnel to incur financial obligations in the local community and then move on without paying. If this type of behavior happens, it reflects on the client company's image, since the general public is unaware that the shopper is not a permanent employee. Then, too, other kinds of social misbehavior on the part of contract engineering personnel can create the same sort of poor image.

Ineffective workers. As if irresponsibility were not enough, the user critics have also claimed that many shoppers are ineffective when they actually do work. Most job shops have competent people, but in an emergency a client may have to take the ones who are available—and there are many "poor" or "fair" individuals who are job shopping. One company has voiced the opinion that nine out of ten shoppers are not as good as its permanent employees. A problem in this area appears to be that the shopper/client relationship is impersonal, and shoppers do not respond to client supervision as do permanent employees.

Poor economics. Although, as was mentioned in Chapter 2, a third of the surveyed user firms considered economics to be a major reason for using contract engineering, another third considered economics to be a disadvantage. These firms—even though they continue to use the concept—have suggested that the shoppers' premium pay and the technical risks make for doubtful cost control. This they claim to be particularly true on a small project where a slight overrun is magnified by the high rates of pay. Then, too, if the assignment period is short (two months or less), time that has been spent in orientation cannot be made up.

During conditions of large-scale employment of shoppers, the direct labor base shrinks and overhead rises. One firm that used contract engineering to eliminate layoffs noted that a delicate balance must be maintained between costs and the numbers of permanent and temporary personnel. It is important, they said, to have the largest possible permanent staff without having to lay people off during the low period of the manpower requirement curve.

Drop in permanent employees' morale. Another problem

cited by users is the morale of the permanent employees when contract help is brought in. Some permanent employees resent the contract engineer because his take-home pay is larger and yet he needs their help in doing the job and finding his way around. In these cases, shoppers use poor judgment in disclosing their pay to permanent employees. The captive engineers also know that the shopper has no accountability and will be gone before the company builds to his engineering. Resentment has been known to go so far that supervisors have abused the contract engineers.

In some cases, permanent employees resent the apparently lax ways of the job shopper and fear that he may be a trouble-maker. A large influx of contract personnel may be particularly disrupting to permanent employees, representing changes in the status quo and the existing company social system. Any alterations in standard personnel policies affect the equilibrium of captive personnel.

Such charges as these indicate a lack of smooth working relations and a failure of the permanent staff to accept temporary employees who are brought in to work with them. It would appear that in these cases more time can be lost in bickering than can be gained by the added manpower. Furthermore, any costs saved by the use of temporary help may be lost again if permanent employees resign and increase the turnover rate.

The author's survey of user firms delved into questions related to the seriousness of friction and morale with captive staff, but found little cause for real alarm. For instance, in only one of the 45 surveyed organizations was it necessary to raise the pay rates of captive personnel to offset friction and discontent caused by the captives' working directly with contract technical personnel. Furthermore, only 24 percent of the sampled users actually experienced friction at all—and in these cases, although the incidents of friction proved to be plentiful, they were of a petty nature and not serious enough to degrade morale substantially. Most of the sampled firms headed off the potential problem by training the permanent personnel to understand and appreciate job shoppers in advance of the workers' appearance.

Security problems. A disadvantage of using contract engineering is the problem of security. One aspect of this problem is the necessity of preventing contract engineers from divulging proprietary information to competitors. The user may simply avoid assigning such work to contract personnel, but it will thereby lose the flexibility that is a major benefit of the concept.

It may also be difficult to obtain security clearances for individual contract employees in order for them to be allowed to work on classified government programs. If the individual has previously held the required clearance for a program whose clearing agency is the same as that for the current program, it may take four to six weeks to obtain the reactivated clearance. If the employee has not held an active clearance, it may take two to three months before clearance is obtained. The significance of this disadvantage is that nonclassified work must be found for the contract employees while their clearances are being sought. Also, there is the chance that the clearance may not be granted, and the probability of this happening is greater in the case of contract employees than it is in the case of permanent employees.

Reasons for Nonuse

The reasons for use and the advantages and disadvantages have been presented in describing the role and limitations of contract engineering. It is pertinent also to present the reasons that nonusers of the service have given for making their decision against the concept.

Reasons for Nonuse	*Percent of Sample*
1. Rates too high	41
2. Poor image of shops	33
3. Build own technical staff	33
4. Fear of unqualified personnel	30
5. Poor morale of captive employees	26
6. Reluctance to delegate design responsibility	26

	Percent of Sample
7. Insufficient control over job shoppers	18
8. Inability to obtain top management or customer approval	18
9. No fluctuation problems	15
10. Loss of proprietary information	15
11. Severity of on-job requirements	11
12. Risk involved in using small shops	7
13. Overtime promised to own personnel	3

These reasons were obtained from a survey of representative contract engineering firms. Some of the reasons are obviously based on considered judgment, and others on hearsay and ignorance of the advantages of using the contract engineering concept. Since they are essentially the reasons given to salesmen for the companies' not accepting the concept or patronizing the contract firms, it should be recognized that covert reactions may have existed, and that the salesmen received only the overt response. Then, too, since the salesmen were refused, their perception of the situation may have been clouded.

Many of these reasons have been discussed elsewhere, and no attempt will be made here to elaborate them further, except for a comment on the most frequently observed reason of high rates. A review of the cost considerations, which are described in some depth in Chapter 4, will indicate that a refusal to consider use solely on the basis of high labor rates constitutes an obvious lack of situational analysis.

Problems of the Industry

The major problems of an industry that supplies contract engineering services to a variety of other industries appear to be related to both the disadvantages of use and the reasons that nonusers give for avoiding use. Indeed, half the major problems as they are seen by both suppliers and users are reflected in the prior sections of this chapter. The following comments on other problems were obtained from a representative sample of both suppliers and users in the Baltimore–Washington area, and it is believed that this sampling of problems is also representative of the national situation.

Problem as Seen by Supplier	*Percent of Sample*
1. Poor image	45
2. Ethics	41
3. Procurement/personnel department avoidance	21
4. Client bidding practices	21
5. R&D funding availability	17
6. Unqualified personnel	14
7. Severe competition	10
8. High cost of service	7

Problem as Seen by User	*Percent of Sample*
1. Unqualified personnel	55
2. Overclassified personnel	25
3. Misrepresentation	20
4. Lack of interest and loyalty in personnel	18
5. Lack or loss of special knowledge	16
6. Resistance to permanent hiring	13
7. Shop's unconcern for shoppers	11
8. Morale degradation	11
9. Severe competition	9

Contract firms are particularly sensitive about their image as the prime problem; they believe that it has been tarnished by operators who are not technically trained and who select the wrong personnel for clients. They also believe that some clients use the service incorrectly and then receive a bad impression. One contract firm manager volunteered an example of an incorrect approach: A vice-president of a large company patronized a particular job shop that was too small to handle the specific job, and he was provided with unqualified personnel. In other instances, clients may contribute negatively to the shop's image by not allowing enough lead time for proper screening. In still other cases, clients themselves may not know their exact skill or man-hour requirements.

It has been said that dishonesty crops up occasionally on the part of both supplier and user firms. A contract may contain a kickback such as a flat sum per month, or a fixed rate per job shopper per month, directed usually to purchasing agents, sometimes to personnel managers, and occasionally to chief engineers who are interested in augmenting their salaries. It has been alleged, too, that in some instances shoppers have been treated unscrupulously by the shops—where pay, raises granted by the client, and overtime are withheld. Occasionally a shopper has left a company before the work being done is completed, an incident sometimes attributed to a readiness on the part of shops to entice engineers from one another with offers of increased pay.

Some users claim that they have encountered misrepresentation on the part of a job shop firm, and those who have experienced this are reluctant to take at face value the information provided by contract firm salesmen. It is true that an overenthusiastic salesman may oversell both the shop and its personnel, leaving the client to discover—perhaps too late—that the ability of the personnel is one or two levels below classification. One firm stated that it had to review four or five men before it found one that matched qualifications. Another said, "If you want a draftsman, ask for a designer—it appears costly but it gets the job done."

Some shops attempt to deal directly with their clients' engineering management instead of with procurement and personnel specialists. They attempt to avoid the latter because they feel that communication is better with engineering people, where an atmosphere of mutual respect and trust is more readily generated.

Contract firms say that the bidding practice of their clients creates problems for their industry. For one thing, it causes the shops to fight with one another for inexpensive help—who may be incompetent. A contract firm may also underbid a job, knowing full well that it can't produce the requirements for the price, but hoping to increase the price at a later time. Because large users frequently engage a number of contract firms at one time, one group of shoppers may discover that another group is being paid more, and may attempt to be hired by the high-rate firm. A bidding problem associated with small R&D clients involves their tendency to deal with many different contract firms, even though they may want only two shoppers at any one time. If they used just a few shops, they would probably experience better service.

Quality of Contracted Services

Most of the lists of limitations associated with contract engineering mention a dissatisfaction with the qualifications of temporary personnel. The remainder of this chapter will attempt to identify job shoppers' levels of qualification, as compared with permanent engineering staffs in the Baltimore–Washington area in 1968–69. How representative this regional sampling is of the quality level for other parts of the country is unknown. According to several users and suppliers in the sample, the region studied included a greater proportion of subqualified contract help than did other regions—the New England area, for example, where a greater proportion of degreed engineers were reported to be job shopping. If this is true, the reader is invited to provide his own adjustments to the quality level described here, so that they may reflect his geographical area.

The individual differences of people is a major factor to consider when evaluating technical personnel. The word "qualified," unless defined in context, is intangible. There have been many cases in practice where an engineer with experience and proper academic degrees should have been highly qualified for an assignment but was found to be a novice when compared with a seemingly less qualified individual who may have possessed greater native intelligence and more pertinent experience. Therefore, when this book identifies a person as being more qualified than another, it means that he is *apparently* more qualified, as determined by his academic degrees and years of experience.

Shoppers were compared with permanent engineering employees with respect to apparent levels of technical qualification. If the levels were the same, it could be said that all classifications of job shoppers were utilized to the same level, as represented by the captive population; but if they were significantly different, it could indicate a highly professional use of job shoppers, or a concentration of subprofessional use.

The measure used in comparing the qualification level of personnel was a nondimensional quality index rating derived from a mathematical model shown in the appendix. The magnitude of the calculated index rating represented the average (weighted) quality of the permanent technical staff of a company unit or of the group of shoppers rented to a company unit. Weights were arbitrarily assigned to represent value differences between levels of education and experience possessed by the technical staff being considered. In addition, opinions were sought from engineering managers of user firms. These opinions related general thinking on the quality of rented personnel versus captive personnel, and allowed for the inclusion of intangible characteristics such as creativity.

Statistical testing of the selected samples indicated that contract firms reported their personnel to be significantly more qualified than they were reported to be by users. The calculated mean quality index numbers were 3.05 and 2.72, respectively. Qualitatively, the suppliers indicated that their average employee was college trained (nongraduate) and had experience; while the users indicated that he was a high school graduate with experience, and that he had less college training. This result was expected. A contract firm that believes in its product can be expected to be favorably biased; conversely, user firms may tend to enhance the qualifications of their own personnel even if it means underevaluating contract personnel.

When the quality of job shoppers was compared with the quality of permanent personnel, the permanent personnel were found to be significantly more qualified than the contract personnel, regardless of the source of data. The calculated mean quality index for permanent personnel was 3.51, which represents personnel most of whom are college trained and half of whom are graduates with many years of experience. In a review of index numbers for contract personnel, it is seen that the difference in average talent between job shoppers and captive employees is pronounced.

Analysis of variance results showed that the size of the client company had no effect on the quality of contract personnel who were used. It could also be said that the suppliers did not discriminate on the basis of company size in sending talent to clients. The same variance results showed, however, that company size did have an effect on the quality of *permanent* personnel used. This was particularly significant when large companies were compared with small companies. Such a result is not surprising, since large companies generally have more resources to expend on maintaining a highly qualified technical staff.

The results of an opinion survey conducted among user managers are illustrated in Exhibits 2 and 3. When the creativity figures were compared with the qualifications figures, it

Exhibit 2

Technical Qualifications of Contract Personnel Compared with Permanent Personnel (Opinion Survey)

	Large Companies		*Medium Companies*		*Small Companies*		*All Companies*	
Relative Qualifications of Contract Personnel	*No. of Replies*	*Percent*	*No. of Replies*	*Percent*	*No. of Replies*	*Percent*	*No. of Replies*	*Percent*
More	1	5	2	14	2	14	5	11
Equal	8	42	6	43	4	29	18	38
Less	10	53	6	43	8	57	24	51
Total Replies	19	100	14	100	14	100	47	100

Exhibit 3
Creativity of Contract Personnel
Compared with Permanent Personnel
(Opinion Survey)

Relative Creativity of Contract Personnel	*Large Companies*		*Medium Companies*		*Small Companies*		*All Companies*	
	No. of Re- plies	*Per- cent*	*No. of Re- plies*	*Per- cent*	*No. of Re- plies*	*Per- cent*	*No. of Re- plies*	*Per- cent*
More	0	0	2	14	1	7	3	6
Equal	5	28	5	36	2	13	12	26
Less	13	72	7	50	12	80	32	68
Total Replies	18	100	14	100	15	100	47	100

was discovered that users generally tended to be more negative about contract personnel's creativity than about their qualifications. In other words, they believed that for equal qualifications, the creativity of permanent personnel would be greater than that of job shoppers. A comparison of opinions on the basis of client size indicated that managers of large and small companies considered contract personnel to be much less creative than did managers of medium-size companies.

A United States patent rate study was also conducted as a measure of creativity. It used contract firm data for job shoppers, because user data resulted in a zero rate for them. Despite this, the rates for permanent employees from all sizes of companies were significantly higher, when tested at the 5 percent level, than for the rates of contract personnel. Contract personnel were therefore shown to be inferior in creativity when patents were used as a measure.

4

The Question of Cost

THE available literature, as well as the discussions in this book, indicates that discriminating use of temporary technical help will provide tangible advantages. Many companies translate the advantages into cost savings, while others insist that the use of contract personnel costs more than non-usage. Some managers in the latter category believe that the small additional cost is well compensated for by their companies' ability to produce and deliver a product on time. The point to be emphasized here is that generalizations about the comparative expense of contract engineering should be viewed with caution. Whether it saves an organization money or costs it money depends upon the specific situation, the many variables involved, and an intelligent application of the service.

When hidden and intangible labor costs are considered in detail, temporary help can often be supplied by a contract firm for less money than it would cost to hire a permanent employee. The hidden costs were emphasized in a study conducted in 1964 by the New York management consulting firm of Stevenson, Jordan, and Harrison. This study, entitled "The Economics of Using a Temporary Help Service," evaluated the hidden costs of permanent employees as being 33.8 percent of base salary.[1] Since the average contract firm's markup is reported as being about 1.35 times payroll, it is evident that in many cases the cost of direct hiring can be extensive.

Exhibit 4 presents a cost breakdown listing hidden cost items that are conventionally described as fringe benefits. This table reflects national averages for the fringe benefits listed as a percentage of payroll. The unemployment compensation part of "legal payments" amounts to 3.1 percent on a national average basis, although in some states it is considerably more. In the District of Columbia and the states of Maryland and Virginia, unemployment compensation is a percentage of the first $3,000 per year earned by the individual, and it varies from a minimum rate of 0.5 percent for a good employment record to a maximum of 4.7 percent for a poor record, including the federal portion of 0.4 percent.

Stevenson, Jordan, and Harrison also reported that since the daily workload of any one permanent employee varies from month to month, the possible savings of using a temporary employee can be calculated. They recommended multiplying the actual number of hours worked by a permanent employee during the month in question by the hourly cost of an equivalent leased employee and then subtracting this value from the sum of the permanent employee's average monthly salary and fringe benefit cost. The result is the potential savings, whose value is based on the true number of hours worked rather than those for which man-hours were expended in "mark time" situations.

Some other intangible cost considerations were suggested by Robert S. Eckley. Concerned with the high cost of a layoff policy, he recommended that novel and positive approaches be taken to stabilize the work force. He emphasized that the biggest problem in activating novel approaches is changing the attitude of the managers. "The visceral response of many manufacturing managers to the experiments mentioned and to the

[1] Marion M. Whalen, "Renting People Is Good Business," *Credit and Financial Management*, February 1965, pp. 12–15.

Exhibit 4

The Cost of Fringe Benefits, 1964

Cost Items	*Percent of Payroll (National Average)*
1. *Legally Required Payments*	7.825
Social Security	
Unemployment compensation, state	
Unemployment compensation, federal	
Workmen's compensation	
Other	
2. *Pensions and Payments*	10.5
Insurance	
Contributions to private unemployment funds	
Separation and termination allowances	
Discounts on goods and services purchased	
Meals furnished	
3. *Paid Items*	3.0
Rest periods	
Lunch periods	
Washup time	
Travel time	
Clothes change time	
Get-ready time	
4. *Paid Time-Not-Worked Items*	10.4
Vacations or payment in lieu thereof	
Holidays	
Sick leave	
National or state guard duty	
Jury duty	
5. *Miscellaneous*	2.1
Profit sharing	
Christmas bonus or special bonus	
Service and suggestion awards	
Educational expense	
Total	33.825

objective of employment stability is that it wouldn't work here."[2] This same attitude is not uncommon among engineering managers where job shopping or other stabilizing forces are being considered for use.

Mr. Eckley said that when a layoff is imminent, the apparent cost of savings to the company of the wages involved is the key, but seldom is the cost of layoff itself subtracted from the so-called savings. He noted that the following layoff costs should be considered:

1. Layoffs initiate higher unemployment compensation taxes on the basis of experience rating systems in almost all states. Furthermore, the taxes may continue at the higher rate for several years even if employment goes back up.
2. When employees are shuffled because of downgrading resulting from the layoffs, and then the cycle turns upward with expanded production and upgrading, the total proceedings are accompanied by a temporary "confusion factor" loss estimated to be about 2.5 percent of the annual payroll.
3. If the country's history since World War II is used as a guide, former production levels are exceeded in one to three years after a recession. This means sizable hiring and training costs.

These considerations make the layoff savings questionable if the company is planning to stay in business. Furthermore, these cost components emphasize the need to consider contract engineering as a device to eliminate layoffs in the technical staff.

Differentials, based on accumulated 1967 data, which existed between the costs of using permanent technical employees and the costs of using contract technical employees were investigated by the author for the Baltimore–Washington area. These differentials are presented to serve only as a guide for managers in

[2] Robert S. Eckley, "Company Action to Stabilize Employment," *Harvard Business Review*, July–August 1966, pp. 51–61.

determining cost relationships; a fuller analysis of their development and implication is presented in the appendix.

The reader is cautioned to view the results of this analysis objectively for several reasons:

1. Costs apply only to one geographical area for the year 1967. The reader who may be interested in a different area and time period should modify the results by considering applicable factors—or by conducting his own study.
2. Not all cost elements were included for analysis.
3. A few industries (chemical, for example) were represented by very small sample sizes.

The cost elements included in the analysis for comparison were average permanent employee regular wages; fringe benefits; overtime costs; costs of hiring permanent technical employees during 1967 apportioned among all permanent employees; contracted hourly costs of job shoppers on both regular time and overtime; and the contract markup when data on shoppers came from job shop firms. Intangible costs were generally excluded from this study and left to the judgment of the engineering manager contemplating the use of contract engineering. Cost elements that were not included were training and adjustment costs for new captive and contract employees; confusion-factor costs resulting from previous layoffs; costs arising from slack periods without layoffs or transfers; overhead costs associated with both the physical plant and the services provided to captive and contract employees; costs of the technical inefficiency of individual personnel, and so on.

A simplifying assumption was made for purposes of analysis: that a linear relationship existed between the cost of technical services and the quality of the services available. This assumption applied to the use of both permanent and contract employees. It inferred that two engineers of equal competence and experience levels were paid equally well within the industry or related industries in which they worked. It did not assume equal pay for equal competence from one industry to another where the pay scales may have been different.

The above assumption was further limited by realities. Although one would like to have believed that equal competence in an industry resulted in equal pay, it was known that the pay of an individual technical employee depended, in the final analysis, upon how well he marketed himself. In the long run and on the average, however, more competence resulted in better wages—but individual differences contributed to a widespread scatter of randomly selected sample points.

The analysis of the tangible costs of using contract technical services as compared with the services of permanent technical employees is summarized and presented with average values in Exhibit 5. A selected group of regression lines for all the various industries and for shopper and captive employees are shown plotted in Exhibit 6. This type of plot permitted comparison of the unit costs of shopper and captive employees when their qualifications were equivalent.

Linear regression analysis of tangible costs per hour of the use of contract and permanent employees on varying skill levels in various industries indicated that the unit costs of using contract employees, based upon equivalent quality, were considerably higher than for permanent employees.

All the basic regression levels except one were positively sloped, which indicated that some degree of regression between cost and quality level existed. The one slightly negatively sloped line (not shown in Exhibit 6) represented shopper employees in the aerospace–ordnance–federal agency group. It indicated that this particular industry group, the one most experienced in using shoppers, was obtaining highly qualified shoppers at almost the same cost per hour as permanent, highly qualified personnel. The large difference in cost for subprofessionals such as draftsmen and technicians indicated that these types were in high demand and that the industry had to pay the price for their services when they were needed.

Regression analysis of costs based on quality for a combined industry grouping consisting of federal agencies and companies

Exhibit 5
Comparison of Hourly Costs: Captive and Shopper Employees

Type of Employee	*Industry Group Represented*	*Sample Size, n*	*Average Cost, $/hour*	*Average Quality Index No.*
Captive*	Electronic–electrical	18	6.681	3.535
Shopper	Electronic–electrical	20	7.760	2.547
Captive*	Aerospace, ordnance, and government agency	15	6.895	3.556
Shopper	Aerospace, ordnance, and government agency	11	7.465	3.02
Captive*	Heavy and machinery	5	5.588	3.162
Captive*	Architects and engineers	3	4.71	3.30
Shopper	Heavy, machinery, architects, and engineers	8	6.255	2.682
Shopper*	All industries–data from user companies	41	7.34	2.718
Shopper*	All industries–data from supplier companies	23	8.00	3.02
Captive*	Electronic, electrical, aerospace, ordnance, government agency–combined	33	6.793	3.544

*Regression curves plotted in Exhibit 6.

Exhibit 5 (continued)

Type of Employee	*Industry Group Represented*	*Sample Size, n*	*Average Cost, $/hour*	*Average Quality Index No.*
Shopper	Electronic, electrical, aerospace, ordnance, government agency—combined	31	7.665	2.719
Captive	Electronic, electrical, aerospace, ordnance, government agency—combined—large size	14	6.893	3.741
Captive	Combined—medium size	10	7.003	3.359
Captive	Combined—small size	9	6.348	3.444
Shopper	Electronic, electrical, aerospace, ordnance, government agency—combined—large size	12	8.395	2.952
Shopper	Combined—medium size	9	7.560	2.511
Shopper	Combined—small size	10	6.895	2.625
Captive	Chemical	2	7.045	4.065

Exhibit 6

Regression of Cost on Quality for Captive and Shopper Employees

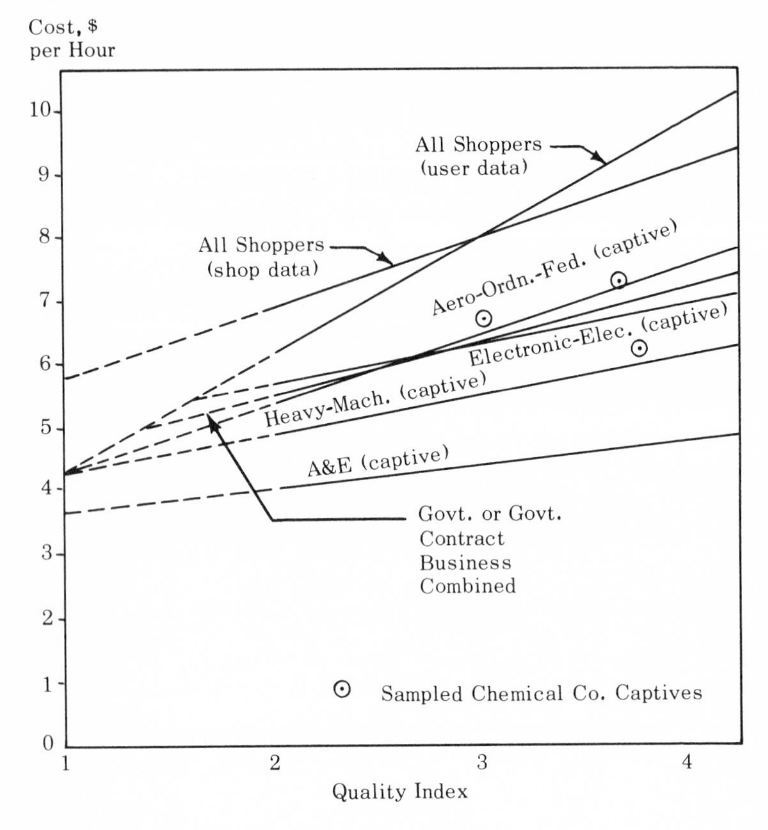

that service the government—classified according to organization size—showed no significant differences for permanent employees. However, for contract employees, the small companies had a very small positive slope as compared with the slope for medium and large companies. This indicated that small companies had obtained higher qualified people at lower cost. Perhaps contract firms have reduced the rates for them specifically, in order to encourage them to utilize the services.

When the cost element of wages alone is considered, analysis indicates that the average wages of contract personnel and captive personnel in the Baltimore–Washington area were 6.5 percent in favor of the shopper—$6.78 and $6.40 per hour, respectively, for contract and permanent personnel. These values included fringe benefits. The average quality index values for these same personnel, based on user data, were 2.72 and 3.51, respectively. This classified the average contract personnel as subprofessionals, and average captive personnel as professionals. But the subprofessionals earned $0.38 per hour more as individuals than did the professionals. An implication of these figures is that most captive technical personnel can increase their hourly remuneration by job shopping.

5

Contract Engineering: The Future

THE contract engineering concept is valuable if its use is based upon proper weighing of alternative and situational factors by a discriminating manager. The manager contemplating its use must consider alternatives with good judgment, because

1. The unit cost for equivalent quality may be high. If timing is not carefully watched, use of the concept can offset flexibility and other advantages on which the decision for use is based.
2. The average contract employee in some geographical areas may be substandard in qualification and creativity. Thus time on the job must be carefully planned by user supervision, and care in selection must precede acceptance of individuals from the sources of supply.

Users of the service regard the concept as a valuable approach to staffing, because they can meet peak manpower requirements quickly and then destaff with equal dispatch. They are not required to maintain a large permanent staff that in the long run would cost them money through inactivity of personnel when the peak has passed. Furthermore, they have the opportunity to obtain certain scarce skills that may not be resident on the permanent staff but are needed urgently at a specific time.

A Philosophy and Approach for Potential Users

The approaches available to the manager in acquiring contract technical services personnel are numerous. However, the most likely source of supply is the technical services industry. Thus when a management decision is made to use job shop help, the decision should include a tentative selection of one or more contract firms with which to do business. There are an adequate number of contract firms from which to choose, many of which are conscientious firms that are in business to provide substantial service to clients as well as to make a profit.

In order to obtain the greatest value from their association with contract firms, potential clients need to be both honest with with themselves and selective. To minimize improper practices, engineering management should communicate directly with and select contract firms, with assistance from the personnel department. This should not be a procurement department function, since highly skilled employees are involved and not materiel.

Mr. Leon N. Skan, former president of the National Technical Services Association (NTSA), told the National Association of Purchasing Agents just what major points should be considered in the selection of a technical services firm with proper qualifications. The message was well directed to buyers of materiel, and should also be useful to engineering and personnel management whose major function involves technical staffing.

1. *Analysis of reputation.* Determine from other companies whether the technical service firm under consideration has the needed talent, a good record, and sound man-

agement. Also check references for participation in other programs and for the competence of the people in the programs.

2. *Analysis of the firm's approach.* Be sure that the firm has executives who show a complete understanding of technical manpower problems and individual capabilities of people needed to handle the job properly.
3. *Analysis of costs.* It may be very unwise to pick the firm with the lowest bid, because sometimes a firm that analyzes the job correctly will recommend more skilled men who command a higher rate but who can complete the job in less time.[1]

The industry that Mr. Skan represents is new and has grown rapidly. In a national survey conducted by NTSA in 1967, confirmed by the author's local survey, the mean founding year for job shop companies was 1957. The industry is maturing and many of those who were in it for the "fast buck" have fallen by the wayside. Furthermore, with less money available for R&D from the government, and with the general economy in a recessed condition, marginal contract firms have essentially been eliminated—but not all of them. It is still necessary to discriminate among suppliers to obtain satisfactory job performance from rented talent.

One of the contract firms represented in the author's survey recommended a specific technique for contracting with job shops, advising against the purchase-order approach and recommending a blanket one-year agreement with clients, based on careful manpower planning. Each user was advised to patronize three reputable shops for its own protection, and to use a "sole source" agreement with each job shop to obtain better rates. The sole-source agreement is one that lasts for two weeks; if the supplier is not supplying, the opportunity then reverts to another vendor.

[1] "Leon N. Skan Addresses 53rd Conference of the National Association of Purchasing Agents," *The PD News,* June 17, 1968, pp. 8–12.

The Future Predicted

The average age of technical services firms in this country is 13 years. Much has occurred during these years to shape up the concept and expand the use of contract engineering. But what lies ahead? What will the industry look like after another 13 years? What differences are anticipated?

Almost half the 75 surveyed suppliers and users visualized that there would be more growth and demand for contract technical services in the years to come. One user said: "The future offers exciting growth potential. We can't see the leveling point. Industry is learning to use the concept and hasn't yet learned all the possibilities." Another user noted that: "Growth will continue because a regular company during the growth period of its life needs the shops; it can't afford the luxury of a heavy recruiting campaign." Many other using companies believed that the present use of the concept would expand at an increasing rate—or at least keep pace with the gross national product. On the other hand, supplier companies anticipated growth in the next ten years to vary from 20 to 400 percent over present conditions. One supplier qualified his growth prediction to limit it to firms that were providing sound services to clients.

Even though 50 percent of the users predicted growth, there were almost another 30 percent of them that predicted a stabilization or decline. A user company producing electronic business machines volunteered that growth would not keep up with the economy, because until recently growth has been disproportionately fast. Another user company based its prediction of a smaller demand upon increased automation through the use of electronic computers and cathode-ray display devices that permit quick data recall and immediate calculation by the designer. "In ten years we will talk to computers and have a decreased demand for shoppers. Furthermore, automatic drafting machines are already current and will replace rented draftsmen," it said. Other users noted that the immediate future looked dim because of a lack of available funds and the continuance of the hostilities in Vietnam. They believed that a depressed period following the end of the military action would cause an employment slowdown —after which the job shop business would grow again. One of them noted that, with a leveled-off economy, the marginally qualified shoppers would not be acceptable. Contract business would then decline rapidly, cleaning out the remaining "deadwood" shops.

The suppliers that visualized a decline in growth were few in number. However, they believed that the going would be difficult in the next few years unless the shops offered services in hard-core-unemployable training and housing, and in other poverty programs in keeping with the national interests.

A major prediction made by 25 percent of both the users and the suppliers related to improvement in job shop practices. Several users and suppliers suggested that the shops would be forced to perform a better personnel screening job. Others noted that the shops would take more interest in having employees continue their education and remain up to date with engineering changes and trends, so that both the quality of performance and the contract firm's image would be improved. With better practices, the shops would appear to be more professional and would attract a better supply of talent.

It is significant to note that 25 percent of the users and 30 percent of the suppliers predicted that the overall engineering community would become accustomed to contract employee usage, and that more companies would use it to avoid layoffs, the costs of recruitment, and poor community images. The renting of everything needed in the office, factory, and home, they said, has become a way of life and will include technical personnel. Thus the use of the concept will spread to encompass a greater number of nondefense-oriented industries. More organizations will be using it as a first choice and fewer as a last resort. Several users visualized that job shops might dominate the market for engineers and draftsmen. This would contribute to direct hiring difficulties and stimulate some industries to rent shoppers more frequently.

The "crystal ball" has helped to foresee more responsible work assignments for rented help. Surveyed users believed that both specialized teams and high-grade specialists would be rented more frequently to solve immediate problems. The suppliers concurred that departmental development teams, tailored to the needs of the client, would be used. These would include management as well as technical services. The suppliers also foresaw more project responsibility for the work, with a resultant decrease in captive/contract employee intermingling.

Fifty percent of the suppliers of contract technical services predicted a basic change in the structure of their industry. They believed, on the basis of current trends, that the small shops would be squeezed out of existence. This would be accomplished through acquisition, growth of the good shops, and the increasing difficulty for small shops to enter the business. One supplier thought that the reduction in total numbers of shops would follow an approximate 4 to 1 ratio (one quarter to remain). Another volunteered that in 20 years there would be just 20 large companies with many branches and with powerful lobbies to encourage their use.

In summary, the relatively new technical services industry, according to users and suppliers, will continue over the long run to grow and prosper. It will accomplish this growth, not at the expense of the companies that it serves, but with their assistance, because the services rendered will be worth the costs. The conditions that must be met to assure this bright future are:

Shop practices

1. Better shop management through technical and business administration qualifications of administrators.
2. Improved screening of personnel and matching of assignments with talents.
3. Development programs for shoppers for continued education and personal growth.
4. More frequent selection of employees for permanency.
5. Encouragement of firm loyalty and participation in servicing clients.
6. Schooling of employees in professional ethics.
7. Elimination of misrepresentation of shop qualities to po-

tential users—and follow-up during use to assure that contract and promises are being met.

User practices

1. Analysis of alternatives.
2. Education in using the services more efficiently.
3. Schooling of captive engineers in the reasons and background for a decision to use the shops—before a contract is implemented.
4. Schooling of captive engineering supervisors in human relations and the behavior that will draw out shopper performance instead of inhibiting it.

It behooves the engineering managers of all industries to remain abreast of the technical services offerings and to know, with justifiable reasons, whether the concept of contract engineering is good for them on the basis of their set of "now" circumstances.

Appendixes

Appendix I

Development and Analysis of the Quality Index

As noted in Chapter 3, the formula developed to provide the quality index is a mathematical model representing the actual situation concerning the overall quality of an engineering staff or group. Factor weights for individual technical people in each category are noted as follows:

$f_1 = 1$ A recent high school graduate.

$f_2 = 2$ A high school graduate with more than five years' experience; or a recent college dropout.

$f_3 = 3$ A recent college graduate; or a nongraduate with some college training and more than five years' experience.

$f_4 = 4$ A college graduate with over five years' experience.

$f_5 = 5$	A holder of a master's degree; or a non-college graduate with patents to his name; or a holder of a state professional engineer's license.
$f_6 = 6$	A holder of a master's degree and a professional license; or a doctorate.
$f_7 = 7$	A holder of a doctorate and professional license; or a graduate (B.S.) with patents to his name.

The sequence of these factors and the accompanying weights was assumed to be in an increasing order of importance, and are generally accepted as such by engineering managers when personal individual differences such as personality and native intelligence are not subject to consideration.

The quality index formula represents a weighted index that is dependent upon the percentages of total engineering staff employment of a company unit. It is noted as follows:

$$\overline{PQ} \text{ or } \overline{CQ} = \frac{f_2 p_2 + f_3 p_3 \ldots + f_7 p_7}{100}$$

where:

$\overline{PQ}$ = permanent technical employee quality index rating.
$\overline{CQ}$ = contract technical employee quality index rating.
$f_2, f_3, \ldots, f_7$ = factor weights as previously defined.
$p_2, p_3, \ldots, p_7$ = percentages of technical employees in classifications representative of the factors.

The percentages ($p_1, p_2, \ldots, p_7$) were obtained from the samples of surveyed user and supplier companies. The user companies provided data representative of their own permanent employees identified as $\overline{PQ}$ and their contract employees identified as $\overline{CQ}$. The supplier companies provided supplemental data that permitted calculation of a different $\overline{CQ}$ for each supplier company. This latter information was then compared with the average quality of technical help that users claimed to be renting.

The user firms of contract technical services were specifically asked to estimate the current education and experience levels of their permanent technical employees and of the contract technical services personnel they had employed. Each user company was guided to respond in terms of the percentages of its permanent and rented people who fell into the previously noted factor categories. The supplier firms were asked to respond similarly and to estimate the education and experience levels of their current technical employees working in the Baltimore, Maryland–Washington, D.C., area.

Analysis of the response consisted in calculating the individual $\overline{PQ}$s for each user company in the sample of companies from the accumulated percentage data. The $\overline{CQ}$s from both supplier and user data were also calculated for each sample unit.

Testing Equality of Populations

The two sample populations of $\overline{CQ}$s were compared to see whether they could be classified as being from the same universe. The comparison was made by calculating the mean values of $\overline{CQ}$ and the variance for each sample, and then testing to see whether the population variances were the same. This was followed by a test to see whether the population means were the same.

In order to test the equality of variances, the F statistic was constructed by using unbiased estimates of the population variances calculated from the two samples as follows:

$$F = \frac{\hat{\sigma}^2_{\overline{CQ}} \text{ (from users)}}{\hat{\sigma}^2_{\overline{CQ}} \text{ (from suppliers)}}$$

where the numerator has the largest variance.

Tables of the F distribution at the 5 percent level of significance using $(n_u - 1)$ and $(n_s - 1)$ degrees of freedom provided the critical point for a one-tailed test. Since the calculated value of F was less than the critical value, the hypothesis that the variances were equal was accepted.

Since the variances proved to be equal, the population means were tested using the t statistic and the F distribution. The t statistic is as follows:

$$t^2 = \frac{(X_u - X_s)^2 (n_u n_s)(n_u + n_s - 2)}{(n_u S_u^2 + n_s S_s^2)(n_u + n_s)}$$

where:

n_u = sample size of $\overline{CQ}$ from user data.
n_s = sample size of $\overline{CQ}$ from supplier data.
S_u^2 = variance of $\overline{CQ}$ sample from user data.
S_s^2 = variance of $\overline{CQ}$ sample from supplier data.
X_u = sample mean of $\overline{CQ}$ from user data.
X_s = sample mean of $\overline{CQ}$ from supplier data.

A one-tailed F test at the 5 percent level of significance for 1 and $(n_u + n_s - 2)$ degrees of freedom was made to locate the critical value for t^2. If the calculated value had been less than the critical value, the hypothesis of equal population means and hence equal populations would have been accepted. Since it was more, the null hypothesis was rejected, and the populations were declared unequal.*

It was necessary to calculate the sample mean and the estimated variance of the $\overline{PQ}$s so that the quality of captive engineers could be compared with the contract variety. The comparison was made with the $\overline{CQ}$ population. The previous analysis indicated that there were two apparent $\overline{CQ}$ populations, and the comparison with the captive quality was made with both indicated contract employee populations.

The technique used for comparing $\overline{PQ}$ samples with $\overline{CQ}$ samples was identical to the prior analysis using the F and t^2 statistics and the tables of the F distribution for critical values at the 5 percent level of significance. A summary of the results of the foregoing analysis is shown in Table A.

* Taro Yamane, *Statistics: An Introductory Analysis*, 2d ed. New York: Harper and Row, 1967, pp. 651–659.

Table A
Comparison of Quality Index Populations for Contract and Captive Personnel

Statistical Parameters	*Results of Population Comparison* $\overline{CQ}$ (*user data*) *vs.* $\overline{CQ}$ (*supplier*)	$\overline{PQ}$ (*user data*) *vs.* $\overline{CQ}$ (*supplier*)	$\overline{PQ}$ (*user data*)$_u$ *vs.* $\overline{CQ}$ (*user data*)$_s$
$n_u - 1 =$ degrees of freedom	40	44	44
$n_S - 1 =$ degrees of freedom	26	26	40
$n_u + n_S - 2 =$ degrees of freedom	66	68	84
$\overline{X_u}$ = mean of sample	2.720	3.512	3.512
$\overline{X_S}$ = mean of sample	3.050	3.050	2.720
S_u^2 = variance of sample	0.311	0.272	0.272
S_S^2 = variance of sample	0.304	0.304	0.311
$\hat{\sigma}_u^2$ = estimate of population variance	0.318	0.278	0.278
$\hat{\sigma}_s^2$ = estimate of population variance	0.315	0.315	0.318
F statistic $= \frac{\sigma^2 \max}{\sigma^2 \min}$	1.01	1.13	1.145
$F^{df_1}_{df_2}$ = critical point at $\alpha = .05$	$F\frac{40}{26} = 1.85$	$F\frac{26}{44} = 1.75$	$F\frac{40}{44} = 1.66$
Significant	No	No	No
$H : \sigma_u^2 = \sigma_s^2$	Accept	Accept	Accept
$(\overline{X_u} - \overline{X_S})$	0.330	0.462	0.792
t^2 statistic	5.59	12.35	45.20

Table A (continued)

Statistical Parameters	Results of Population Comparison		
	$\overline{CQ}$ (user data) vs. $\overline{CQ}$ (supplier)	$\overline{PQ}$ (user data) vs. $\overline{CQ}$ (supplier)	$\overline{PQ}$ (user data)$_u$ vs. $\overline{CQ}$ (user data)$_s$
$F' n_u + n_s - 2 =$ critical point when $\alpha = .05$	3.98	3.97	3.96
Significant	Yes	Yes	Yes
$H_1 : \mu_u = \mu_S$	Reject	Reject	Reject
$H_2 : \mu_u < \mu_S$	Accept	Reject	Reject
$H_3 : \mu_u > \mu_S$	Reject	Accept	Accept
Populations are:	Different	Different	Different

Notes: Subscript (u) = user data
Subscript (s) = supplier data
(df) = degrees of freedom

Analysis of Variance for Effect of Company Size

The effect of company size on the quality of contract technical talent used and the effect of company size on the quality of permanent technical staff hired involved two separate analyses utilizing the same method. The analysis, which was conducted to determine the effect of size on $\overline{CQ}$, is described here in detail. The identical approach was used to determine the effect of size on $\overline{PQ}$.

The values of individual company $\overline{CQs}$ as defined from data supplied by the user were segregated into the classifications of "large" (L), "medium" (M), and "small" (S). The mean value of $\overline{CQ}$ for each sample of company classification was then calculated together with the estimate of the classification (population) variance. It was then necessary to test to see whether the classification variances were equal. Since they were equal, an analysis of variance was performed to see whether the classification means were equal. If the means proved unequal, then company size did make a significant difference, and it was necessary to determine which specific classifications were unlike each other.

The method used to determine the equality of variances was identical to the previous analysis described for equality of the two $\overline{CQ}$ population variances. The comparison of variance was accomplished in two steps by first assuming $\sigma_L^2 = \sigma_S^2$, and then assuming $\sigma_L^2 = \sigma_M^2$. (See Table B.) Since all three popula-

Table B

Company Size Effects on Quality Index Number: Test of Equality of Variances

Statistical Parameters	$\overline{CQ}$ *Large vs. Small*	$\overline{CQ}$ *Large vs. Medium*	$\overline{PQ}$ *Large vs. Small*	$\overline{PQ}$ *Large vs. Medium*
$n_L - 1$	14	14	18	18
$n_M - 1$ or $n_S - 1$	14	10	13	11
σ_L^2	0.286	0.286	0.332	0.332
σ_M^2 or $_S$	0.335	0.358	0.298	0.168
F statistic $\frac{\sigma^2_{max}}{\sigma^2_{min}}$	1.17	1.27	1.112	1.975
F with $(n_L - 1)$ and $(n - 1)$df, critical point when $\alpha = .05$	2.48	2.86	2.48	2.68
Significant	No	No	No	No
$H : \sigma_L^2 = \sigma_M^2$ or $_S$	Accept	Accept	Accept	Accept

Notes: L = Large
M = Medium
S = Small

tion variances proved to be equal at the 5 percent level, testing was continued using an analysis of variance procedure.

The model used for analysis of variance was the fixed effects, randomized, one-way classification model.* The approach hypothesized that all population means were equal; that is, $\mu_L = \mu_M = \mu_S$. Three treatments represented by the three com-

* Yamane, pp. 677–694.

pany sizes were used, and sample size (n_i) for each treatment was taken as the value for the classification of size which was smallest in the survey. The other two samples were selected by the random number table from the data of the companies sur-

Table C

Analysis of Variance: Company Size Effects on Quality of Contract Personnel

Source	*Sum of Squares*	*Degrees of Freedom*	*Mean Square*
Between	1.0	2	$S_A^1 = 0.500$
Error	10.3	30	$S_E^1 = 0.343$
Total	11.3	32	

F statistic $= \frac{S_A^1}{S_E^1} = \frac{0.500}{0.343} = 1.46$

F distribution critical point for 2 and 30 degrees of freedom at 5% level significance = 3.29

$H_1 : \mu_L = \mu_M = \mu_S$ is accepted.

Table D

Analysis of Variance: Company Size Effects on Quality Index of Permanent Personnel

Source	*Sum of Squares*	*Degrees of Freedom*	*Mean Square*
Between	2.575	2	$S_A^1 = 1.287$
Error	8.713	33	$S_E^1 = 0.264$
Total	11.288	35	

F statistic $= \frac{1.287}{0.264} = 4.87$

F distribution critical point for 2 and 33 degrees of freedom at 5% level of significance = 3.29

$H_1 : \mu_L = \mu_M = \mu_S$ is rejected.

H_2 : The population means are not all equal is accepted.

Table E

Results of Pairwise Tests: Company Size Effects on Quality Index of Permanent Personnel

Table of $\lvert X_i - X_j \rvert$	X_j / X_i	*Large* 8.99	*Medium* 3.89	*Small* 2.85
Large	8.99	. . .		
Medium	3.89	5.10	. . .	
Small	2.85	6.14*	1.04	. . .

$$|(X_i - X_j)| \geqslant \sqrt{2n_o S^1{}_E F^1{}_{33}\,(.05)^1}$$

$$\geqslant \sqrt{2(12)(0.264)(4.14)} = 5.12$$

*The large and small sizes are significantly different at the 5% level of significance. They are from different populations. The large–medium and small–medium sizes are equivalent populations.

veyed. The purpose here was to select equal sample sizes for the analysis of variance. A table of observations and a table of squared observations were prepared, from which the total sum of squares, the between-treatment sum of squares, and the error (within) sum of squares were calculated. The degrees of freedom for between of $(a - 1) = 2$, and the degrees of freedom for the error $(n - 2)$ were used to calculate the mean squares. From this information the F statistic was constructed as follows:

$$F_o = \frac{\text{Mean square for between}}{\text{Mean square for error}}$$

The F distribution was used to determine the critical value at the 5 percent level of significance for degrees of freedom of two and $n - 2$, which were compared with F_o above. If F_o was less than the critical value, the hypothesis that $\mu_L = \mu_M = \mu_S$ was accepted. If F_o was more than the critical value, tests were conducted to determine which classification was different from the others. This test utilized the following formula for significance at the 5 percent level:

$$|(X_i - X_j)| \geqslant \sqrt{2n_o S_E{}^1 F^1{}_{n\,-\,a}(.05)}$$

where:

$|(X_i - X_j)|$ = absolute value difference between the sum of the observations for a treatment and a like sum for another treatment.

n_o = number of observations for each treatment.

$S_E{}^1$ = mean square for the error term.

$F^1{}_{n\,-\,a}(.05)$ = critical value from the F distribution for 1 and $(n - a)$ degrees of freedom.

The results of this analysis are summarized in Tables C, D, and E.

Appendix 2

Cost Analysis

THE methodology used in determining the approximate tangible costs of using contract technical services, with the limitations set forth in Chapter 4 kept clearly in mind, was to combine product quality information from the prior section of this appendix and wage information (including fringe benefits) together with supplemental cost data obtained from surveying the user, so that average hourly costs of engineering in the year 1967 for each sampled user could be plotted against a corresponding engineering quality index. Regression lines for captive employee costs and contract employee costs were constructed using the paired cost and quality data. These lines were then compared with each other.

The relative average rates of pay for a 40-hour work week of captive and contract personnel were found first by obtaining related data from the user and supplier surveys. The pertinent questions asked of the supplier during the survey were

1. What is the current average wage of your individual technical employee working in the Baltimore, Maryland–Washington, D.C., area?

2. What is the current average weekly per diem paid?
3. Please state your average wage for the individual services you offer.
4. Please indicate the fringe benefits you supply to each employee: Insurance ______, Pension ______, Vacation days ______, Holidays ______, Unemployment tax ______, Other ______.

The applicable questions asked of the user during the survey were

1. Please state the average regular wage in dollars per hour paid by you during 1967 for the services of your permanent engineering and technical employees.
2. Please state the average fringe benefits paid to your permanent engineering and technical employees in terms of dollars per hour or percent of regular wage.

The topic of wage rates was sensitive both for the contract firms and for the user firms, since both types were competing within their respective industries. During initial testing of user firm questions, it was found that a detailed breakdown of average wages by skill classification was very difficult to acquire. Therefore, the requests were changed to average total wages for the technical staff. Reluctance, then, was not severe. However, contract firms reacted differently; reluctance was observed, but compliance depended on the philosophy of the company. Some preferred to give classification wage rates; others, a lumped average; and still others, no information at all.

The average contract firm's wage was determined by the following steps:

1. Wage data from contract firms, providing a breakdown of wages according to services rendered, were screened and grouped into three skill-wage areas to which percentages of the firms' technical staff were assigned. These skill-wage areas were contract engineer wage, contract technical writer wage, and an

average wage for combined contract draftsman, contract illustrator, and contract technician.

The last three subprofessional skills were grouped because a review of raw data indicated that the wages were always very close. Furthermore, almost all of them were included in the education and experience category below that of college training.

2. Percentages of contract engineers in the workforce of a firm were assumed to be those with a B.S. degree and five years' experience; those with patents and professional licenses; and those with advanced degrees. The percentages of technical writers were assumed to fall in the experience and education category between the engineers and the subprofessionals already mentioned. This was the classification reserved for recent college graduates and for experienced people with college training.

3. The hourly wages representative of the three skill-wage areas were multiplied separately by the appropriate percentage of the contracted staff, and these values were added to arrive at the average wage of contracted personnel for a particular firm.

Example for contract firm A:

Contract engineer wage = \$10.00 per hour
Contract technical writer wage = \$6.50 per hour
Average grouped wage of contract personnel:

Draftsman	=	\$5.00
Illustrator	=	4.00
Technician	=	5.00
Subtotal		\$14.00

Average: $\frac{14.00}{3} = \$4.67$ per hour

Engineers in technical workforce = 80 percent
Technical writers in workforce = 10 percent
Draftsmen, illustrators, and technicians in workforce = 10 percent

Average wage = (10.00)(.8) + (6.50)(.1) + (4.67)(.1)
= \$9.12 per hour

4. Data from firms not supplying these details, but providing an average technical wage value identified as not representing the inclusion of in-house wages, were added to the calculated wage values, and a mean wage for the total sample was determined.

The fringe benefits paid to job shoppers were estimated by calculating the average number of holidays, vacation days, and sick days reported by suppliers, and then converting the days into dollars per year per man by multiplying the equivalent missed hours by the base hourly wage. The portion of fringe benefits associated with Social Security and unemployment tax (both federal and state) was estimated to be 7 percent of the first $3,000 earned per year per employee. The 7 percent included 4.4 percent for Social Security payments and 2.6 percent for unemployment tax. The latter value was estimated as the middle of the rate range of 0.5 percent for a good employment record to 4.7 percent for a poor one in the Maryland–District of Columbia–Virginia area. Assuming that a reasonable number of contract firms paid all or part of a job shopper's hospitalization and accident insurance (with some life), a flat estimate of $0.10 per hour was considered as the insurance part of the total fringe benefit. The value of pensions, paid education, bonuses, and so on were estimated from the survey data.

The approximate value of the fringe benefit "package" paid to job shoppers on a per hour basis was calculated by adding all the individual fringe value items together to obtain a total fringe payment for the year. This payment was then divided by the number of straight-time hours worked per year to arrive at an hourly value for fringe benefits. The hours worked per year were based upon 2,080 hours less vacation, holidays, and sick days. A summary of fringe benefit calculations is shown in Table F. Per diem payments to job shoppers were uncommon in the Baltimore–Washington area and were not considered as a fringe benefit. When it was necessary to pay this charge, however, its value was generally accepted as $56 per week. Captive employee wage and fringe benefit average values were calculated directly from the user survey data. The comparison of wage information for the two types of technical employees consisted in contrasting average wages with and without the amount of the respective fringe benefits. The results are shown in Table G.

Table F
Summary of Average Fringe Benefit Values for Contract Technical Personnel

Benefit	*Value*
Holidays: (6.38) (8) (6.18), $ per year per employee*	$315
Vacation: (5.82) (8) (6.18), $ per year per employee*	210
Sick pay: (2) (8) (6.18), $ per year per employee*	99
FICA, and federal and state unemployment tax, $ per year per employee	210
Insurance @ $4 per week	208
Total, $ per year per employee	$1,042

*[(days per year) (hours per day) (hourly rate)]

Table G
Average Remuneration to Contract Technical Personnel and Captive Personnel: A Comparison

Comparative Information	*Contract Technical Personnel*	*Captive Technical Personnel*
Sum of company average wages, $/hr.	142.20	225.89
Sample size, number of firms	23	43
Sample mean wage, $/hr.	6.18	5.25
Standard deviation wage, $/hr.	1.11	0.77
Fringe benefits, $/hr.	0.60	1.15
Fringe benefits, percent of base pay	9.7	21.9
Standard deviation, $/hr. (fringe)	. . .	0.38
Total of basic wage and fringe, $/hr. average	6.78	6.40
Quality index rating, average	3.05[a] 2.72[b]	3.512

Notes: [a] From supplier.
[b] From user.

The following supplemental questions were asked the sampled users to provide a basis for cost analysis:

1. How many permanent engineering and technical staff members are employed (clerical help omitted)?
2. How many permanent technical employees have you hired during 1967?
3. Please estimate your average hiring costs per permanent technical employee including advertising, personnel department or agency services, moving expenses, per diem, etc. $________ per man.
4. Please indicate the average percentage of straight-time hours devoted to overtime by your permanent technical staff during 1967.
5. What was the average overtime pay factor for your permanent staff (e.g., 1.00, 1.25, 1.50, . . .)?
6. What was your average cost, in dollars per straight-time hour, for services of contract technical personnel leased from all contract firms utilized in 1967 (if none were used in 1967, please estimate the average value for those used within the past two years)?
7. What was your average overtime cost factor for leased personnel?
8. What were the average number of hours over forty per week worked by your leased technical employees?

The average dollar cost per engineering man-hour used for the year 1967 for permanent employees was calculated for each user company in the sample which supplied the pertinent data. This cost was represented by the symbol $\overline{PE}$. It excluded overhead charges associated with the physical plant and intangible costs. Also, the average dollar cost per engineering man hour used for the year 1967 paid to contract firms for services was calculated for each user company in the sample that supplied the required data. This cost was the billing cost of the contract services and was represented by the symbol $\overline{CE}$. Both costs, $\overline{PE}$ and $\overline{CE}$, were calculated for each user company in a systematic, step-by-step procedure utilizing a table format with each company representing a column heading. The row headings are shown in Table H. They illustrate the required steps used in the analysis to arrive at hourly cost results.

A separate set of $\overline{CE}$ costs were calcuated from the average wage rates for contract technical services as determined from the supplier survey and contrasted with the results of the user survey. The costs were calculated assuming an average Baltimore–Washington area markup factor of 1.30. This factor was conservatively chosen to be below the national average of 1.35 because of concentrated competition in the area. The purpose of this second set of data was for verification of the overall contract rate structure. It did not lend itself to categorization by industry or by company size.

On the basis of calculated values of $\overline{PE}$ and $\overline{CE}$ together with paired quality index values of $\overline{PQ}$ and $\overline{CQ}$ obtained from the previous analysis described in this appendix, the data were further segregated into like industry groupings. Regression lines were constructed for each group.

The method of linear regression analysis employed in the study was the method of least squares. $\overline{PE}$ and $\overline{CE}$ were the random variables plotted on the Y-axis, and $\overline{PQ}$ and $\overline{CQ}$ were the nonrandom variables plotted on the X-axis. A line using the model format, $Y = bX + a$, was constructed from the data so that:*

$$b = \text{slope} = \frac{n\,\Sigma XY - \Sigma X\,\Sigma Y}{n\,\Sigma X^2 - (\Sigma X)^2}$$

$$a = Y\text{–intercept} = \frac{\Sigma Y}{n} - b\frac{\Sigma X}{n}$$

$n =$ sample size
$X = \overline{PQ}$ or $\overline{CQ}$
$Y = \overline{PE}$ or $\overline{CE}$

Construction of these regression lines and calculation of the corresponding sample correlation coefficients were based upon a

* Yamane, p. 383.

Table H

Cost Analysis Format for Calculation of $\overline{PE}$ and $\overline{CE}$

Row Number	*Analysis Step*
1	Permanent technical staff, number
2	Cost of hiring a professional man, dollars
3	Professionals hired in 1967, number
4	Cost of professional hires, 1967, (2) × (3), dollars
5	Cost of hiring a subprofessional man, dollars
6	Subprofessionals hired in 1967, number
7	Cost of subprofessional hires, 1967, (5) × (6), dollars
8	Total cost of technical hires, 1967, (4) + (7), dollars
9	Hiring apportioned over permanent staff, (8) ÷ (1), dollars per employee
*10	Hiring apportioned, $ per hour worked per employee, (9) ÷ 1,864 hours
11	Overtime per permanent employee per year, hours
12	Permanent employee regular wage, dollars per hour
13	Permanent employee fringe benefit, dollars per hour
14	Permanent employee overtime rate factor
15	Permanent employee overtime rate, (14) × (12), dollars per hour
16	Yearly overtime costs per man, (11) × (15), dollars
*17	Yearly regular time costs per man, 1,864 [(12) + (13) + (10)], dollars
*18	$\overline{PE}$ = [(16) + (17)] ÷ [1,864 + (11)], dollars per hour per man
19	Contract employee cost, dollars per week per man (40 hrs.)
20	Contract employee overtime costs, overtime rate × hours per week worked, dollars per week per man
21	$\overline{CE}$ = [(19) + (20)] ÷ [hours per week worked], dollars per hour per man

*1,864 hrs. = actual regular hours per year worked = 52 weeks × 40 hrs. less 216 hrs. for holidays, vacations, and sickness.

population where the type of distribution of Y on X was unknown. Therefore, the Type I population noted by Taro Yamane† and his corresponding methods of analysis were utilized. The correlation coefficient was regarded as a measure of the closeness of fit of the regression lines to the points, a measure of the degree of linearity of the scatter of the points, and a measure of the amount of improvement (reducing the total error) due to the regression line. The coefficient of regression of "one" represented a perfect fit of the sample points to the line. On the basis of the type of data represented in this study, a coefficient of approximately one-half was considered to be good. This was achieved in 9 of 17 regression lines.

Computation procedures for the correlation coefficient, r, and the adjusted coefficient, r (used where the number of degrees of freedom was small), are noted as follows:‡

Total error = Unexplained error + explained error

$$\Sigma(Y - \overline{Y})^2 = \Sigma(Y - Y_c)^2 + \Sigma(Y_c - \overline{Y})^2$$

where:

$$S_{YY} = \Sigma(Y - \overline{Y})^2 = \text{sum of squares, total.}$$

$$S_E = \Sigma(Y - Y_c)^2 = \text{sum of squares, unexplained error.}$$

$$S_R = \Sigma(Y_c - \overline{Y})^2 = \text{sum of squares, explained error.}$$

$$r = \sqrt{\frac{S_R}{S_{YY}}}$$

$$r = \sqrt{1 - \frac{\dfrac{\Sigma(Y - Y_c)^2}{n-2}}{\dfrac{\Sigma(Y - \overline{Y})^2}{n-1}}}$$

and:

$Y = \overline{\text{PE}}$ or $\overline{\text{CE}}$ data point.

$\overline{Y}$ = average value of the Y points.

Y_c = calculated value from developed regression line.

† Yamane, p. 371.

‡ Yamane, pp. 391–403.

Bibliography

Consulting Engineers Council of the USA. Personal interview with Donald A. Buzzell, Executive Director, Washington, D.C. November 25, 1968.

Drexel University. Personal interview with LeRoy A. Brothers, Provost, Philadelphia, Pennsylvania. June 9, 1968.

Eckley, Robert S. "Company Action to Stabilize Employment." *Harvard Business Review*, July–August 1966, pp. 51–61.

Ergler, Paul C. "A Study of Contract Engineering as Used in Technical Manpower Management." unpublished doctoral dissertation, School of Government and Business Administration, The George Washington University, 1969.

Fouquet, David. "New Job Gypsies Boast Skill, Top Pay." *The Washington Post*, May 1, 1966.

Gold, Martin. "Job Shop Revisited." *Electronic News*, July 18, 1966.

"Job Shops: They're Big Business." *Engineering News Record*, August 10, 1967, pp. 15–17.

"Leon N. Skan Addresses 53rd Conference of the National Association of Purchasing Agents." *The PD News*, June 17, 1968, pp. 8–12.

Milani, Ernest J. "The Contract Engineer." *Mechanical Engineering,* September 1967, pp. 52–53.

———. "The New Breed of Gypsy Technicians." *Personnel,* November–December 1967, pp. 56–60.

National Technical Services Association. Letter and report of 1967 survey from Sheldon J. Hauk, Executive Director, Chicago, Illinois. February 6, 1968.

Pacey, Margaret D. "Money Making Form: Providers of Temporary Help Are Working to Regain It." *Barron's,* March 4, 1968, p. 11.

Rumpf, Frank H. "How Temporary Help Boosts Our Economy." *Office,* January 1967, p. 92.

Whalen, Marion M. "Renting People Is Good Business." *Credit and Financial Management,* February 1965, pp. 12–15.

Yamane, Taro. *Statistics; An Introductory Analysis.* 2d ed. New York: Harper & Row, 1967.

About the Author

PAUL C. ERGLER has been associated with the aerospace industry since 1941, following his graduation in mechanical engineering from Drexel Institute of Technology in Philadelphia. He was first employed by the Glenn L. Martin Company in Baltimore as an airplane stress engineer; and then by the Martin Marietta Corporation as a structural design specialist, engineering section chief, program technical director, engineering department manager, and quality engineering manager. His most recent work has been in market research with the aerospace group staff.

Dr. Ergler holds an M.S. degree in engineering management from Drexel and a doctorate in business administration from The George Washington University (1969). While conducting his dissertation research in 1967–1969, he taught management courses in the MBA program at Loyola College in Baltimore. He returned to Loyola as Associate Professor of Business Administration in 1970.